EVO-DEVO OF
CHILD GROWTH

EVO-DEVO OF CHILD GROWTH

Treatise on Child Growth and Human Evolution

ZE'EV HOCHBERG

WILEY-BLACKWELL

A JOHN WILEY & SONS, INC., PUBLICATION

Published by John Wiley & Sons, Inc., Hoboken, New Jersey
Published simultaneously in Canada

For general information on our other products and services or for technical support, please contact our Customer Care Department within the United States at (800) 762-2974, outside the United States at (317) 572-3993 or fax (317) 572-4002.

Wiley also publishes its books in a variety of electronic formats. Some content that appears in print may not be available in electronic formats. For more information about Wiley products, visit our web site at www.wiley.com.

Library of Congress Cataloging-in-Publication Data:

Hochberg, Z.
  Evo-devo of child growth : treatise on child growth and human evolution / Zeev Hochberg. – 1
    p. cm.
  Includes index.
  ISBN 978-1-118-02716-5 (hardback)
1. Child development. 2. Children–Growth. 3. Human evolution. I. Title.
  RJ131.H595 2012
  618.92–dc23

                                                                              2011037214

Printed in the United States of America

oBook ISBN: 978-1-118-15615-5
ePDF ISBN: 978-1-118-15612-4
ePub ISBN: 978-1-118-15614-8
eMobi ISBN: 978-1-118-15613-1

10 9 8 7 6 5 4 3 2 1

The following scientists contributed notes:

Andreas Androutsellis-Theotokis: Technical University, Dresden, Carl Gustav Carus Medical University, Medical Clinic III, Carl Gustav Carus University, Dresden University of Technology, Germany and European Brain Research Institute, Rome, Italy

Stefan Bornstein: Technical University, Dresden, Carl Gustav Carus Medical University, Medical Clinic III, Carl Gustav Carus University, Dresden University of Technology, Germany

George Chrousos: Division of Endocrinology, Metabolism and Diabetes, First Department of Pediatrics, Athens University, Children's Hospital Aghia Sophia, Athens, Greece; John Kluge Chair in Technology and Society, Library of Congress, Washington, DC, USA

Ken Ong: MRC Epidemiology Unit, Cambridge, and MRC Unit for Lifelong Health and Ageing, London, UK

Moshe Szyf: Department of Pharmacology and Therapeutics, Sackler Program for Epigenetics & Psychobiology, McGill University, Montreal, Quebec, Canada

Alan Templeton: Department of Biology, Washington University, St. Louis, Missouri, USA

CONTENTS

PREFACE

The study of human growth has focused over the years mostly on descriptive statistics and the investigation of hormonal mechanisms. In the past few years, expressions from evolutionary biology such as "thrifty gene" and "thrifty phenotype," "developmental programming" and "adaptive responses" have found their way into the jargon of scientists and practitioners in the field of child growth. This book reviews the ways in which evolutionary thought clarifies child growth and maturation.

In the summer of 2003, during a routine search of articles in preparation for a research project, I came across the theory of life history. The review by Barry Bogin on the evolution from two preadult life history stages in early hominids to four in *Homo sapiens* (Bogin 2002) provided me with a tool to logically analyze child growth from a different angle than medical. After then reading Hillard S. Kaplan and Jane B. Lancaster's *An Evolutionary and Ecological Analysis of Human Fertility, Mating Patterns, and Parental Investment* (Kaplan and Lancaster 2003), I realized that evolutionary biology could be integrated with the medical approach to child growth and maturation, and came up with new research questions that such integration generates. I learned that evolutionary thinking offered a new and powerful insight into the work that I had been doing both in my clinic and my research for the past three decades. It provided a continuing supply of new questions posed from a different perspective leading to alternative rationalization of the medical phenomena and offered new directions in my research of child growth. It occurred to me that Johan Karlberg's infancy–childhood–puberty model for child growth (Karlberg 1989), which had been a working tool in my and other clinics for years, is much in line with the life-history theory, and I began to explore the correlations between the life-history approach and the infancy–childhood–puberty model. The findings were astounding for me, and during a hike in the summits of the Galilee Mountains, an important concept of this treatise crystallized: The transitions between life-history stages—from infancy to childhood, then to juvenility and adolescence—are unique periods available for evolutionary adaptive adjustment to the environment.

Further exploration of descriptive auxology and endocrine mechanisms proved they were "proximate" tools for understanding the sequence and mechanisms of these transition periods. I realized that along with bipedalism and our evolutionary focus on brain development and language, childhood, as a life-history stage, is the essence of humanity. I also learned that an additional life history stage juvenility, which is much discussed in evolutionary biology, had to be redefined in clinical terms. Most life-history investigators and clinicians in the field of child growth used to classify this period, if at all, as the years from pubarche to the onset of puberty. Analysis of the data required that I change this definition. In the study of adolescence, I discovered that our young do not become adults when adolescence completes, and defined a new life-history stage that I call youth, from the last trimester of adolescence to age 24.

This treatise contains some original work on child growth, and since the conclusions at which I arrived after drawing up a rough draft appeared interesting to me, I thought that they might interest others. Many of the views presented here may be claimed to be speculative, and some will no doubt prove erroneous, but I have in every case given the reasons that have led me to one view rather than to another.

My personal conceptions of child growth and maturation are based on my own learning, observations, and experience as a pediatric endocrinologist. Yet the preparation of a book such as this inevitably owes a great debt to the many researchers who have investigated child growth, human evolution, and life-history theory. Their names are too numerous to list here, and they are detailed in the reference section. The presentation of growth charts as first and second derivatives in relation to life-history stages is a novelty of this book, and I hope that both clinical practitioners and scientists of child growth will find them useful. The data used for these calculations come from two sources: the published U.S. National Center for Health Statistics (NCHS), 2000 Centers for Disease Control (CDC) Growth Charts, and Gerver and de Bruin's 1996 reference manual *Pediatric Morphometrics* (Gerver and deBruin 1996).

I wish to thank my colleagues who wrote Notes for this book: Alan Templeton, Ken Ong, George Chrousos, Stefan Borenstein, Andreas Androutsellis-Theotokis, and Moshe Szyf. I owe a special debt of gratitude to my colleague and friend Kerstin Albertsson-Wikland from Göteborg, Sweden, who hosted me as an adjunct professor in her department while we jointly developed the concept of delayed infancy–childhood transition (DICT) (Hochberg and Albertsson-Wikland 2008). I am thankful to those colleagues who contributed to some of the work on evolutionary concept we have jointly done: Drs. Dov Tiosano, Ron Shaoul, Alina German, Aneta Gawlik, Robert Walker, Hasan Eideh, Michael Shmoish and Bjorn Jonsson, Yonatan Crispel, and Oren Katz. I have used here parts of a review article I wrote with Drs. Feil, Constancia, Fraga, Junien, Carel, Boileau, Le Bouc, Deal, Lillycrop, Scharfmann, Sheppard, Skinner, Szyf, Waterland, Waxman, Whitelaw, Ong, and Albertsson-Wikland (Hochberg, Feil et al. 2011). I express my appreciation and thanks to two colleagues who read the manuscript and gave me much useful advice and help: Avigdor Beiles, emeritus population geneticist and evolutionist from Haifa University, and Israel Hershkovitz, physical anthropologist from Tel Aviv University.

I hope that the ideas presented here—a different approach to child growth and maturation—find a wide and understanding audience.

ZE'EV HOCHBERG
April 2011

1

INTRODUCTION

A. EVOLUTIONARY THINKING IN MEDICINE

Evolutionary medicine is a relatively new discipline at the junction where evolutionary insights clarify medical observations and where the latter offers not only new insights but also research opportunities in evolutionary biology. Yet, the gaps are still wide between evolutionary biology and clinical medicine. Here are some concepts as to where the two vary.

Medical practitioners and medical scientists view organisms as machines whose design has been optimized by engineers to provide good health. The evolutionary perspective asks why those mechanisms are the way they are. It views organisms, instead, as compromises between traits shaped by natural selection to maximize reproduction, not health or the good of the species (helping it to avoid extinction) (Stearns, Nesse et al. 2010). They are the product of inevitable trade-offs. Thus, compromises that increase disease resistance often have costs, and some variations that increase susceptibility have their benefits. Evolutionists like to make the distinction between "proximate," for mechanistic explanation, and "ultimate," for evolutionary explanation, which in medicine is often misunderstood as teleological—design of final causes that exist in nature. The synthesis of the two approaches, which is used throughout this book, is that proximate mechanisms evolved through interactions with the environment to shape phenotypes.

Evolutionists place special emphasis on the concept that selection acts on phenotypes, not on genes, and in a similar manner, patients are phenotypes. Doctors do not treat genes; they treat traits that are influenced by genes and their expression. Whereas much of the theory of life history has been developed and tested without reference to the genetic underpinnings of the trait (admittedly, Charles Darwin

Evo-Devo of Child Growth: Treatise on Child Growth and Human Evolution, First Edition.
Ze'ev Hochberg.
© 2012 Wiley-Blackwell. Published 2012 by John Wiley & Sons, Inc.

knew nothing about genes), genomics may respond to the question of how a trade-off that is expressed at the phenotypic level is manifested or modulated at the level of genetic regulation.

The notion that common heritable diseases are caused by a few defective genes is usually incorrect. Rather, many genetic variants interact with environments and other genes during development to influence disease phenotypes (Stearns, Nesse et al. 2010). Genome-wide association studies have failed to explain more than a few percent of the variations in a trait. Such studies have shown time and again that most of the genetic variance of a trait is contributed by mutations at low frequency in the population, and the effects of rare mutations tend to be much larger than those of common mutation; mutations that have strong effects on fitness are likely to be rare in populations and hence difficult to detect; and mutations that are easy to detect have small effects on disease.

Another interface between biological evolution and clinical medicine is the recognition of cultural effects on disease. Population evolution is much slower than cultural change, and diseases arise often from the mismatch of our bodies to modern environments.

Two examples illustrate how evolution has provided important insights to medicine: (1) in the explanation of aging and (2) why humans have more cancers than other species. Both aging and cancer are not adaptations, but by-products of selection for reproductive performance earlier in life. We now live for two decades and more of post-reproductive years that are relatively indiscernible to natural selection; we practice our unique sexuality with countless nonreproductive cycles and contraceptives; likewise, we are not adapted to our rich diet.

B. EVO-DEVO

Evolutionary development biology (evo-devo) addresses the issues of how developmental systems have evolved and probes the consequences of these historically established systems for organismal evolution (Muller 2007). Research in evo-devo has formed around comparative embryology and morphology, evolutionary developmental genetics, and experimental epigenetics. In that respect, it considers the interactions of both microevolutionary processes—changes in traits and gene frequencies resulting from selection and drift in each generation at the populations level—and macroevolutionary processes in deep time perceived in comparisons among species and with fossil evidence and phylogenetic lineages. Micro- and macroevolution explain why species and populations are the way they are, but they do not explain individuals (Stearns and Koella 2008). Understanding individuals requires adding considerations of development: the interaction of genes and environment at each stage of life history, a combination that is now referred to as "evo-devo." This treatise takes evo-devo into postnatal life, the realm of clinical medicine, and more particularly, the physiology and pathology of child growth and maturation. Under the evo-devo concept, child growth is a developmental process, and we no longer speak of genetic versus environmental effects and "nature or nurture"; all traits are products of both.

Human populations are usually thought to be poor candidates for the study of basic questions about the evolution and maintenance of fitness traits; the effects of culture are profound, and environments are variable and far different from those in which the species evolved. Humans, however, are the most investigated species, and clinical observations offer an immense opportunity for the study of evolution. Moreover, special cultural conditions offer something like a natural situation and are used in this treatise to understand the role of the environments on child growth.

Plasticity in developmental programming has evolved in order to provide the best chances of survival and reproductive success to the organism under changing environments (Hochberg, Feil et al. 2011). Environmental conditions that are experienced in early life can profoundly influence human biology and long-term health. Developmental origins of health and disease and life-history transitions are purported to use placental, nutritional, and endocrine cues for setting long-term biological, mental, and behavioral strategies in response to local ecological and/or social conditions. The window of developmental plasticity extends from preconception to early childhood and involves epigenetic responses to environmental changes, which exert their effects during life-history phase transitions. These epigenetic responses influence development, cell- and tissue-specific gene expression, and sexual dimorphism, and, in exceptional cases, can be transmitted transgenerationally. Translational epigenetic research in child health is a reiterative process that ranges from research in the basic sciences, preclinical research, and pediatric clinical research. Identifying the epigenetic consequences of fetal programming creates potential applications in clinical practice: the development of epigenetic biomarkers for early diagnosis of disease, the ability to identify susceptible individuals at risk for adult diseases, and the development of novel preventive and curative measures that are based on diet and/or novel epigenetic drugs.

In November 1859, Charles Darwin published the first edition of *On the Origin of Species by Means of Natural Selection, or the Preservation of Favoured Races in the Struggle for Life*. The second edition was published two months later, in January 1860. Chapter 4 of the *Origins* includes the following paragraph, which is as valid today as it was then:

> If during the long course of ages and under varying conditions of life, organic beings vary at all in the several parts of their organisation, and I think this cannot be disputed; if there be, owing to the high geometrical powers of increase of each species, at some age, season, or year, a severe struggle for life, and this certainly cannot be disputed; then, considering the infinite complexity of the relations of all organic beings to each other and to their conditions of existence, causing an infinite diversity in structure, constitution, and habits, to be advantageous to them, I think it would be a most extraordinary fact if no variation ever had occurred useful to each being's own welfare, in the same way as so many variations have occurred useful to man. But if variations useful to any organic being do occur, assuredly individuals thus characterised will have the best chance of being preserved in the struggle for life; and from the strong principle of inheritance they will tend to produce offspring similarly characterised. This principle of preservation, I have called, for the sake of brevity, Natural Selection. Natural selection, on the principle of qualities being inherited at corresponding ages, can modify the egg, seed, or young, as easily as the adult. Among many animals, sexual selection will give its aid to ordinary selection, by assuring to the most vigorous and best adapted males the greatest number of offspring. Sexual selection will also give characters useful to the males alone, in their struggles with other males.

Ten years later, in 1871, the German philosopher Schopenhauer remarked (as quoted by Darwin):

> The final aim of all love intrigues, be they comic or tragic, is really of more importance than all other ends in human life. What it all turns upon is nothing less than the composition of the next generation. . . . It is not the weal or woe of any one individual, but that of the human race to come, which is here at stake.

C. LIFE-HISTORY THEORY

In simple words, then, the theory of evolution by natural selection claims that the process that causes heritable traits that are helpful for survival and reproduction ("fitness" in the evolutionary jargon—the number of descendents that an organism produces) becomes more common, and harmful traits become rarer. Adaptation occurs through the gradual modification of existing structures. It seemed worthwhile to try to see how far the principles of evolution would throw light on a matter as complex as child growth and maturation.

Variations among species in life history such as growth, maturation, and fertility are extreme. Some species mature within a year of birth and allocate up to 50% of their body mass to reproduction, whereas others take several decades before reproducing and allocate just a small percentage of their body mass to each reproductive episode. Anthropologists attempt to explain how the human species has survived in the face of its unique and apparently impossible life history: a helpless newborn with a short duration of breast-feeding; an extended childhood when the offspring are still dependent for protection and food provision on parental and others' assistance; a juvenile period when they mostly provide for and are able to look after themselves, and when competition with adults for food and space is possible only because the offspring are sexually immature and remain small in size; an energy-costly adolescent growth spurt; delayed reproduction into the third and fourth decades of life; menopause, and a uniquely long postmenopausal life.

Life-history evolutionary theory seeks to understand the factors that produce variations in life stages that are found both among and within species. It is another holistic attempt to integrate a wide range of observations of life in human societies; it encompasses the integration of all fields of biology at the molecular, cellular, and organism levels with social sciences, anthropology, evolutionary biology, and psychology, and more recently, clinical medicine. Using a life-history approach, we consider the ways that evolution has worked upon these life stages to produce the adaptations of a society way of life to its environment.

In the case of hominids, it is best understood in the context of biological rationale and cultural expressions as a solution to an ecological problem posed by the environment and subject to constraints intrinsic to humans (Smith 1992). Among the questions posed in life history research are: Why are organisms small or large? Why do they mature early or late? Why do they have few or many offspring? Why do they have a short or a long life span? Why do they grow old and die at a particular age? This treatise is an attempt to use life-history theory in the understanding of child growth and maturation in a broad evolutionary perspective with a special emphasis on the clinical aspects of this theory.

Two essential assumptions of life-history theory are: (1) there are set measures of fitness (a combination of survivorship and reproductive rate; individuals with higher fitness propagate more genes to future generations) that are maximized by natural selection, and (2) these are often associated by trade-offs among traits that limit the adaptive potential of a population concurrently or at a later time. Thus, species that maximize life-history traits, such as fertility, typically cannot simultaneously maximize survival, and in the growth domain, species that maximize offspring size cannot maximize offspring number at the same time. Survival is affected among other by investment in immune function and adipose deposits, whereas body size is achieved by, among other things, the function of the growth hormone–insulin-like growth factor-I axis. The latter stimulates growth size, while at the same time depletes adipose depots and suppresses immune function and affects survival. Indeed, transgenic animal studies have shown that excess growth hormone shortens the life span and growth hormone deficiency prolongs it. Thus, a trade-off exists between body size and survival.

D. EVOLUTIONARY PERSPECTIVE IN CHILD GROWTH AND MATURATION

Growth and maturation have strong effects on an individual's fitness, because they affect the individual's reproductive potential, schedule, and efficiency (Stearns 2000). When comparing the growth of a human child with that of a cat, or even that of great apes,[1] the pattern difference is obvious (Fig. 1.D.1). The concave pattern

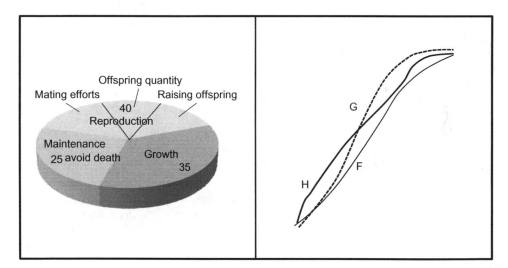

Fig. 1.D.1. The unique human growth pattern and its energetic resources. Left panel: Lifetime resources allocation in Gambian women. Growth requires as much as a third of the lifetime energy allocation. Right panel: Human height pattern (H, bold line) is compared to cat's weight (F, thin line), and gorilla's weight (G, dotted line). The concave pattern of accelerating infantile growth of both cats and apes is in contrast to the convex pattern of decelerating infantile growth in humans. Note also the brisk growth acceleration during human adolescence. Data from Bogin (1999a).

---

[1]Members of the family *Hominins* consist of humans, orangutans, gorillas, and chimpanzees.

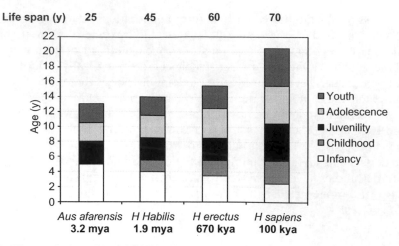

Fig. 1.D.2. The evolution of hominid life history during the first 20 years of life. The existence time is given below [kya (thousands of years ago), mya (millions of years ago)] and the longevity (above). During the evolution of hominids, childhood and adolescence have been added as new life-history stages as compared with apes and presumably the early hominid *Australophithecus afarensis*. The chimpanzee serves as a living representative of the assumed *Australophithecus afarensis* life history. As childhood emerged and prolonged, infancy has gradually cut shorter, and the latest introduced adolescence came at the expense of a shorter juvenility.

(initially slow and accelerating) of nonhuman mammals during their early growth is strikingly different from the convex (initially fast) human growth, followed by a linear growth pattern. This is produced by the uniquely decelerating human infantile growth and the quasi-linear childhood stage.

Whereas several human growth processes are identical to those found in the animal kingdom, hominids' life history is markedly different (Fig. 1.D.2). Humans are born immature, helpless, and defenseless; have a relatively short period of infancy; and are the only species that has a childhood—a biologically and behaviorally distinct and relatively stable growth interval between infancy and the juvenile period that follows. We are also the only species to have true adolescence as a period devoted to puberty and accelerated growth.

Analysis of the height velocity (centimeters per year) in healthy boys and girls of Western European ethnicity, who subsequently become tall or short as adults, reveals the pattern when growth deviates (Fig. 1.D.3). Much of the difference is established early in life during infancy, and some during childhood, where the timing of puberty has minimal impact on ultimate adult height.

The transition from one life-history stage to the next requires a switch mechanism for the onset of the latter, and these switches speak the language of endocrinology, as shown in Fig. 1.D.4 for the sex hormones. Note the rise in sex hormones in early infancy, the so-called mini-puberty, while childhood is characterized by quiescence of sex hormones, followed by a juvenile increase of adrenal androgens, and adolescent increase in gonadotropines and gonadal sex hormones, manifesting as puberty. It is hormones that transduce environmental information to regulate transitions between life-history stages (Hochberg 2009). Indeed, most hormones

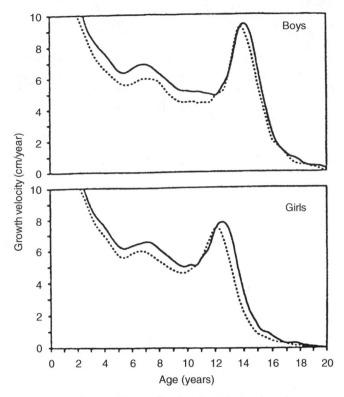

Fig. 1.D.3. Tall and short children. Normative height velocity (centimeters per year) data in healthy boys (upper panel) and girls (lower panel) of Western European ethnicity, who subsequently become tall (solid line) or short (dotted line) as adults. Notice that much of the difference is established early in life during infancy and childhood. Data from Veldhuis, Roemmich et al. (2005).

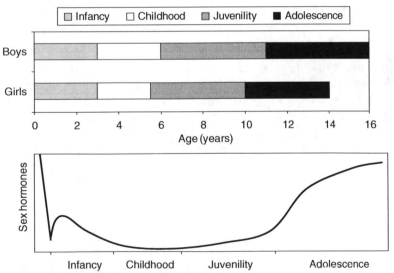

Fig. 1.D.4. Changing sex hormones levels during the first 20 years of human life history. Life-history stages of boys and girls (upper panel) may be defined by sex hormone levels (lower panel). Note the rise in sex hormones in early infancy, the so-called mini-puberty. Childhood is characterized by quiescence of sex hormones, followed by a juvenile increase of adrenal androgens, and adolescent increase in gonadotropine and gonadal sex hormones, manifesting as puberty.

have pleiotropic[2] and often antagonistic effects on a variety of behavioral, physiological, and morphological traits. Multiple hormone mechanisms may have evolved
to activate behavioral and physiological traits at the right time and in the correct
context. When traits are expressed throughout the life cycle, hormones may potentially deactivate them for short periods.

E. CHILD GROWTH AND THE ENVIRONMENT

Charles Darwin, in his *The Descent of Man*, cites the 1869 report of over 1 million
U.S. soldiers, who served in the "late war" and were measured, and the states
in which they were born and reared were recorded. (B.A. Gould, *Investigations
in Military and Anthropology Statistics*, 1869). From this astonishing number of
observations it was proven that local influences of some kind act directly on stature;
and we further learn that "the State where the physical growth has in great measure
taken place, and the State of birth, which indicates the ancestry, seem to exert a
marked influence on the stature."

For instance, it is established "that residence in the Western States, during the
years of growth, tends to produce increase of stature." On the other hand, it is certain
that with sailors, their life delays growth, as shown "by the great difference between
the statures of soldiers and sailors at the ages of seventeen and eighteen years."
Gould endeavored to ascertain the nature of the influences that thus act on stature,
but he arrived only at negative results, namely that they did not relate to climate,
the elevation of the land, soil, nor even "in any controlling degree" to the abundance
or the need of the comforts of life:

> When we compare the differences in stature between the Polynesian chiefs and the
> lower orders within the same islands, or between the inhabitants of the fertile volcanic
> and low barren coral islands of the same ocean (Prichard's 'Physical History of
> Mankind,' 1847).

Darwin continues to quote the same source that there was also a remarkable difference in appearance between the closely allied "Hindoos" inhabiting the Upper
Ganges and Bengal or again between the Fuegians on the eastern and western
shores of their country, where the means of subsistence are very different: "It is
scarcely possible to avoid the conclusion that better food and greater comfort do
influence stature," and:

> Dr. Beddoe has lately proved that, with the inhabitants of Britain, residence in towns
> and certain occupations have a deteriorating influence on height; and he infers that the
> result is to a certain extent inherited, as is likewise the case in the United States. Dr.
> Beddoe further believes that wherever a "race attains its maximum of physical evelop-
> ment, it rises highest in energy and moral vigour" ('Memoirs, Anthropological Society,'
> 1867–69).

With detailed investigations of organisms in their natural environment, one can
determine the potential ecological costs and benefits underlying hormone–
physiology–behavior interactions that, in turn, shed light on their evolution. Such

---

[2]A single factor (or gene) influences multiple phenotypic traits. Consequently, a mutation in the gene
will have a simultaneous effect on all traits.

data also indicate a number of problems, trends, and alternatives for hormonal control mechanisms, and hopefully in the future, a more complete understanding of the common mechanisms underlying behavioral–physiological interactions at the cellular and molecular level. Only then will we be able to predict when and where specific mechanisms of hormone-related interactions operate and how they evolved.

The expression "environment" means different things to different people. For sociologists and psychologists, the environment encompasses social group interactions, family dynamics, and maternal nurturing. For nutritionists, the environment refers to calories, food types, macro- and micronutrients, and dietary supplements, whereas toxicologists think of the environment in terms of water, soil, and air pollutants. However, only a few of these environmental influences have been shown to cause DNA sequence mutations and explain altered gene expression or increases in disease frequency in a particular region (Li, Xiao et al. 2002). Evidence is accumulating that all these very different types of environments are able to alter gene expression and change phenotype by modifying the epigenome (more on epigenetic-mediated plasticity in Chapter 10, Section B). Moreover, when these environmentally induced-epigenetic adaptations occur at crucial stages of life, they can potentially change behavior, disease susceptibility, and survival (Jirtle and Skinner 2007).

Hormonal changes are evident in transitions from one life-history stage to the next. The transition from infancy to childhood is associated with the setting in of the dominance of the growth hormone–insulin-like growth factor-I axis. The transition into juvenility requires the development of androgen-generating adrenal reticularis. The beginning of adolescent-related puberty is evidently a function of the maturation of the hypothalamic–pituitary–gonadal axis. Evolutionary fitness with regard to hormones requires avoiding potential costs and penalties for hormones that are secreted at inappropriate times. Thus, for example, precocious puberty comes at the cost of a loss in ultimate adult height and a "punishment" in the form of psychological and social derisions.

F. HETEROCHRONY AND ALLOMETRY

Evolutionists speak of "allometry" and "heterochrony" in the process of natural selection. The former implies the pattern of covariation among several morphological traits, such as height and head circumference, or between measures of size and shape, such as the growing human cranium and receding mandible. Unlike heterochrony, allometry does not deal with time explicitly.

Heterochrony is the comparison of ontogenetic trajectories between two species representing ancestors and descendants (Gould 1977), such as the shorter infancy in *Homo sapiens* as compared to our ancestors or the ever-growing size of hominids. The dimension of time is therefore an essential part in studies of heterochrony. In the theory of heterochrony, the end point of ontogeny can be altered by the selection of two initial parameters of development: allometric relationship and relative timing for the onset (or offset) of developmental events.

Both heterochrony and allometry are pertinent to understanding child growth and the evolutionary impact and will be used throughout this treatise to understand growth patterns as the child grows through his life-history stages and transitions from infancy to childhood, then to juvenility, and to adolescence and adulthood.

This treatise presents the data and theory of evolutionary predictive adaptive growth-related strategies for transition from one life-history stage to the next and the inherent adaptive plasticity in the heterochrony of such transitions in order to match mostly energy supplies, but also other environmental cues.

G. ADAPTIVE PLASTICITY IN LIFE HISTORY

Phenotypic plasticity is the process by which organisms alter development, physiology and growth, as well as behaviors in response to cues. These responses feature an assessment of both external and internal factors. Physiological sensors compute adaptive trade-offs as a function of energy resources, stress and other signals, and effectors initiate physiological, developmental, and behavioral responses to these determinants. The central nervous system, neuroendocrine circuits, and hormones are critical to growth maturation and development.

I propose the following periods of adaptive plasticity in human's life-history strategies that are related to transitions in life-history stages and child growth:

- Shorter or longer gestation is a plasticity provided as an adaptation for the intrauterine environment, reflected in later metabolic tuning. Thus, among other factors, both preterm and post-term infants are prone to obesity.

- Humans evolved to withstand energy crises by decreasing their body size, and evolutionary short-term adaptations to energy crises utilize a plasticity that modifies the timing of transition from infancy into childhood. A delay in this growth transition, which we refer to as "delayed infancy–childhood transition" (DICT) (see Chapter 4, Section E), has a lifelong impact on stature and is responsible for as many as 50% of children with a normal birth weight and no endocrine disease, who are referred to pediatric endocrine clinics as suffering from "idiopathic" short stature.

- The transition to juvenility is part of a strategy in the transition from the child-hood stage of total dependence on the family and tribe for provision and security until self-supply and a degree of adaptive plasticity is provided and determines body composition. It is associated with specific brain effects of the juvenility hormone dehydroepiandrosterone-sulfate (DHEAS) and changes in body proportions.

- The transition to adolescence entails plasticity in adapting to energy resources, other environmental cues, social needs of adolescence, and the maturation to determine the period of fecundity, fertility, and longevity.

Focusing on child growth, this treatise does not cover the transition from adolescence to adulthood, although it certainly entails a degree of adaptive plasticity. This treatise will further postulate that whereas evolutionary fitness is achieved by the interplay of our entire genome and its expression, life-history stages utilize hormones as a major tool and provide distinct periods of adaptive plasticity in setting in and off of life-history-specific endocrine mechanisms to modify body structures and behaviors. The aim of this text is to discuss the implications of this evolutionary–endocrine–anthropological framework for theory building and new research directions.

2

CHILD GROWTH AND THE THEORY OF LIFE HISTORY

Life history has been defined as the strategic allocation of an organism's energy toward growth, maintenance, reproduction, raising offspring to independence, and avoiding death (Smith and Tompkins 1995). For a mammal, it is, among other factors, the strategy of when to be born, when to be weaned, when to be independent for self-protection and provision, when to accelerate or decelerate growth and when stop growing, when and how often to reproduce, and when to die in the best way as to increase its fitness (Smith 1992). In terms of child growth, life-history models assume that parents make investment decisions that maximize reproductive fitness at the cost of constraints, with a basic rule that energy, effort, and resources that are invested in one direction such as growth cannot be invested in producing more offspring (Stearns 2000). As a consequence, a fundamental trade-off is between size and the fertility rate, with larger mammals having in general fewer offspring (Walker, Gurven et al. 2008).

Relative to other mammals, primates (with humans at its extreme) are slow growing, late reproducing, long lived, and large brained (Charnov 1991). This package is essential to mature to function in the vast variation of environments and the complexity of human society. Other distinctive characteristics of human life histories include:

(1) A period of extreme immaturity after birth, in which the newborn is helpless and defenseless without his mother nearby;

(2) an extended period of dependence of the young beyond breast-feeding age, while the mother becomes pregnant again, resulting in families with multiple dependent children of different ages;

Evo-Devo of Child Growth: Treatise on Child Growth and Human Evolution, First Edition.
Ze'ev Hochberg.
© 2012 Wiley-Blackwell. Published 2012 by John Wiley & Sons, Inc.

(3) multigenerational resource flows and support of reproduction by older post-reproductive individuals;

(4) male support of reproduction through the provisioning of females and their offspring; and

(5) the brain size and its attendant functional abilities are also extreme among humans (Kaplan and Lancaster 2003).

The notion of fitness may be the most important concept in understanding evolution and natural selection. Fitness largely explains the genotypes we see in the world around us. In evolutionary terms, fitness is the quality of a genotype and not that of an individual or a species, although it certainly applies to both. It measures the relative ability of the genotype to reproduce itself. Fitness has a meaning only in comparison to other competing genotypes or specific gene alleles as the extent to which a given genotype or allele is favored by natural selection. Unlike Dawkins's concept of the "selfish gene," modern evolution theories speak about a "fitness landscape" (or "adaptive landscape"), which implies gene interactions among multiple loci—the so-called epistasis.[1] The concepts of fitness, landscape, and epistasis are particularly useful in thinking about complex genetic systems such as those involved in child growth.

Life-history theory predicts that selection will promote efficient physiological and behavioral mechanisms that mediate allocation strategies, for example, by linking life-history traits into trade-offs. Benefits from some function can often only be achieved by incurring costs in other bodily processes at the same or at a later time. Trade-offs result from the need to differentially allocate limited resources to traits like reproduction vs. self-maintenance, or growth vs. longevity, with selection favoring the evolution of optimal allocation mechanism. Since evolution by natural selection works through reproductive fitness, hormone-driven trade-offs play a major role in evolutionary variations. Thus, farmers know that castration of domestic animals diverts resources to somatic growth and metabolic maintenance, thereby increasing growth and protein deposition at the expense of smaller fat deposits.

Variation in life-history traits among species can be extreme, as shown by the variation in age at maturity and energy allocation to reproduction. Some species mature within a year of birth and allocate up to 50% of their lifetime energy to reproduction—the butterfly as an example—whereas others take several decades before reproducing and allocate just a small percentage of their energy to each reproductive episode—the elephant as a case in point. Thus, a vast spectrum of life history stretches from "live fast, remain small, reproduce early, and die young" to "live slow, grow much, reproduce late, and die old" [modified from Charnov (1991)]. The "disposable soma theory" suggests that longevity is determined through the setting of longevity assurance mechanisms so as to provide an optimal compromise between investments in somatic maintenance (including growth and stress resistance) and in reproduction (Kirkwood and Rose 1991). A corollary is that species with low extrinsic mortality (elephants and humans) are predicted to invest relatively more effort in maintenance, resulting in slower intrinsic aging, than species with high extrinsic mortality (the butterfly). The tendency to longevity proved to be

---

[1]Genetic interaction: a gene is modified by one or several genes that assort independently. The gene whose phenotype is expressed is said to be epistatic while the phenotype altered or suppressed is said to be hypostatic.

controlled at the cellular level; cells from long-lived species are indeed more resistant to a variant of stressors than those of short-lived species (Kirkwood, Kapahi et al. 2000).

The human life-history strategy includes "long lived," but presents a special challenge for the general life-history theory. Humans live on a vastly extended timescale compared to most other mammals (twice as long as other apes), and have the longest life span of all terrestrial mammals. Life-history theory needs to explain how humans evolved long gestation and large, yet immature neonates for body weight; an unusually high rate of energy costly postnatal brain growth; an extended period of offspring dependency and slow growth of a helpless young; a relatively brief duration of breast-feeding, yet intense maternal, paternal, and tribal care; delayed reproduction; an adolescent growth acceleration; and menopause with two decades (and today many more) of postmenopausal life (Smith and Tompkins 1995). This unique life history and its intense evolutionary challenges were viewed as pressures that, among others, brought about the evolution of a home base, food sharing, and an elaborate division of labor by gender (Lovejoy 1981), while collaborating in reproductive investment.

The life-history approach adopted here for the study of child growth is a biological and behavioral model of human development, based on a consideration of comparative mammalian biology and primate evolutionary history (Bogin 1999a). It is the product of strategies that *Homo sapiens* use to allocate its energy toward growth, while raising more offspring with a greater survival rate than that of any great ape, and avoiding death for longer than any mammal and into two decades of postmenopausal life (Bogin 2002). The large human brain is an investment with initial costs and later rewards, which coevolved with increased energy allocations to survival (Kaplan and Robson 2002). The ultimate adult size is an evolutionary dilemma for any species; the strategy of intermediate growth stages and the transition between them provides a theory for understanding these strategic objectives that include the length of gestation: when to be born; the length of infancy: when to be weaned; how long to remain a child: when to be independent for self-protection and provision; when to mature for reproduction; and when to cease growth.

Much of the information discussed here was generated from studies of natural fertility societies (Walker, Gurven et al. 2006). "Natural fertility" implies that couples do not practice deliberate fertility control dependent on the number of children they have. In such societies mothers are either pregnant or breast-feed, and will be pregnant again soon after weaning their babies from breast-feeding. These are subsistence-based societies in which most resources are invested as somatic capital in human bodies (i.e., body size and fertility) in contrast to industrial societies, who possess more stored, inherited wealth.

A. LIFE-HISTORY STAGES

Late in the evolution of the *Hominid*,[2] childhood and adolescence have been added as new life-history stages as compared with apes (Fig. 1.D.3). Genetic and anatomical evidence suggests that *Homo sapiens* arose in Africa between 200,000 and

[2]The great apes, including chimpanzees, gorillas, orangutans, and humans.

100,000 years ago, and recent evidence suggests that complex cognition may have appeared between around 164,000 to 75,000 years ago (Marean 2010). Thus, *Homo sapiens* have five prolonged and pronounced postnatal youth life-history stages: infancy, which extends for 30–36 months and ends in natural fertility societies with weaning from breast-feeding, when the mother becomes pregnant again; childhood, which extends for an additional 3–5 years, culminating in a degree of independence for protection and food provision; a juvenile stage, which lasts for 3–4 years and concludes with the readiness for a sexual maturation process; and adolescence, which lasts for 3–5 years, culminating in fecundity (potential reproduction) at an average age of 17 for boys and 15 for girls (Bogin 1999b). The Ju'/hoansi San of the Kalahari Desert (also identified in the literature as the !kung, with the "!" standing for the clicks they use in their language) are hunter-gatherers living in the vicinity of the Namibia/Botswana border. They divide childhood into smaller segments (a "little little girl," a "big little girl," a "little big girl," etc.), even though they have no clear-cut defining boundaries of these stages (Howell 2010).

Yet, the 17-year-old boy and the 15-year-old girl are not fertile in terms of actual reproduction. It will be many more years before they pursue reproduction in modern industrial societies. But even among preindustrial societies, it takes an average of 4 years for fecundity to be realized as fertility (Fig. 2.A.1), later in boys than in girls in most traditional societies (Allal, Sear et al. 2004), and this will be referred to as "youth" (Chapter 8). In an evolutionary fitness perspective, which entails survival of the offspring, the number of children surviving to age 5 years reaches its peak

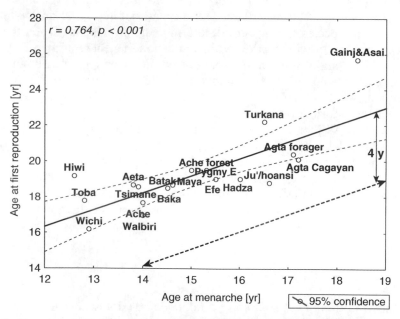

Fig. 2.A.1. Time interval between menarche and the first reproduction. A study of 26 natural fertility societies of hunter-gatherers and pastoralists from around the globe show the 4-year interval from menarche to the first reproduction. Data from Gawlik et al. 2011.

for 18-year-old mothers, with smaller child survival for either younger or older mothers (Allal, Sear et al. 2004).

By reconstruction of indirect evidence, and without certitude, it has been suggested that as late as 3 million to 4 million years ago the early hominid *Australopithecus Afarensis*[3] had merely two postnatal, youth life-history stages: infancy and juvenility (Bogin 1999b). The best known *Australopithecus Afarensis* is "Lucy" from Hadar in Ethiopia, who was considered for many years to be the first hominine, until the 4.4 million-year-old "Ardi," *Ardipithecus ramidus*, was found in the Afar Depression of Ethiopia and shown to be the oldest known skeleton of a putative human ancestor. It is supposed that these hominids had a long infancy of about 6 years (as compared to 5 years of the chimpanzee), but a comparatively short life span, and their first molar teeth erupted in mid-infancy.

With a marked growth of his brain size, 1.9 million years ago *Homo habilis*[4] had to be born less mature to fit into a narrow bipedal maternal pelvis—the so-called obstetrical dilemma (see Chapter 4, Section B) (Washburn 1960; Trevathan 1996). Walking as much as 20 km a day in the open savanna, they also evolved to have longer legs than their tree-dwelling predecessors. *H. habilis* had a shorter period of infancy than his forerunner and developed a new strategic life-history stage: childhood, with its onset defined by weaning from breast-feeding, slowing and stabilization of growth velocity, and dependence on older people for food provision and protection (Smith and Tompkins 1995; Bogin 1999a, 1999b).

While the brain size steadily grew over evolutionary ages (threefold as compared to other apes), newborns became less mature, infancy grew shorter, and childhood was thriving and extending, with the first molar progressively deferred in *Homo sapiens* into the end of childhood and the transition to juvenility. As reconstructions suggest, *Homo sapiens sapiens* (from 100,000 years before present) also introduced adolescence as a distinct life-history stage, with the typical pubertal growth spurt and rapid sexual puberty, deferring the assumption of adulthood in a species that lived longer than his predecessors. The entire assemblage has been extremely successful, and the resulting overpopulation of humans is, for better or for worse, the consequence of that winning strategy. Table 2.1 shows the different life-history stages.

Charles Darwin wrote in 1871 in *The Descent of Man, and Selection in Relation to Sex*, Chapter 6: "On the Affinities and Genealogy of Man":

> Man is liable to numerous, slight, and diversified variations, which are induced by the same general causes, are governed and transmitted in accordance with the same general laws, as in the lower animals. Man has multiplied so rapidly that he has necessarily been exposed to struggle for existence, and consequently to natural selection. He has given rise to many races, some of which differ so much from each other, that they have often been ranked by naturalists as distinct species. His body is constructed on the same homological plan as that of other mammals. He passes through the same stage s of embryological development. He retains many rudimentary and useless structures, which no doubt were once serviceable. Characters occasionally make their

---

[3]The ancestor of the genus *Homo*, which includes the *Homo sapiens*, 3.9 million to 2.9 million years ago.
[4]The species of the genus *Homo* 2.5 million to 1.8 million years ago at the beginning of the Pleistocene era.

TABLE 2.1. Life-History Stages, Growth Milestones and Social Function

Approximate Age	Characteristics of Life-History Stages
0–30 months	**Infancy: Breast-feeding** Feeding by maternal lactation. Rapid and decelerating growth. Rapid growth of the brain. Deciduous dentition. More than 50% of resting metabolic rate is devoted to brain growth and function.
2–7 years	**Childhood: Dependence for Food and Protection** Slowing and stabilization of the growth rate. Immature dentition. Extended family and tribal care. Adiposity rebound. Dependence on older people for food and security. Immature motor control. Cognitive advances.
7–11 years	**Juvenility: Initial Independence** Adrenarche. Find much of own food. Avoid predators. Mid-childhood transition, then slowing growth rate. Compete with adults for food and space.
11–15 years	**Adolescence: Preparation for Adulthood, Nonfertile** Dimorphic secondary sexual characteristics. Growth transition. Peer-grouping. Stature and muscularity are still youth size. Learn and practice adult skills while they are still infertile.

re-appearance in him, which we have reason to believe were possessed by his early progenitors. If the origin of man had been wholly different from that of all other animals, these various appearances would be mere empty deceptions; but such an admission is incredible. These appearances, on the other hand, are intelligible, at least to a large extent, if man is the co-descendant with other mammals of some unknown and lower form.

Kaplan and Lancaster (2003) argued that humans' large brain size and longevity coevolved to respond to their learning-intensive foraging strategies of difficult-to-acquire food resources. These required high levels of knowledge, skill, coordination, and strength, and the attainment of those abilities required an extended learning stage during which reproduction was deferred. This was compensated for by higher reproductive fitness during the adult period, with an intergenerational flow of food from old to young. Thus, by acquiring large brains, humans were selected for lowered mortality rates and greater longevity. Cooperation between men and women that resulted in high levels of male parental investment was promoted when they began specializing in high-quality food, and particularly in hunting. It favored sex-specific specialization in food provision and complementarity between male and female inputs (Kaplan and Lancaster 2003). Food sharing assisted recovery in times of illness and reduced the risk of food shortfalls due to random mortality and fertility. These buffers against mortality also favored a longer childhood, which, as mentioned later, is an essence of the evolutionary success of humans. Cooperation between males and females allowed women to allocate more time to child care, increasing both survival and reproductive rates. The nutritional dependence of multiple young of different ages favored sequential mating with the same individual, since it reduced conflicts between men and women over the allocation of food.

B. TRANSITIONS BETWEEN LIFE-HISTORY STAGES

The developmental life of an organism may be profoundly affected by events during a short critical interval of its life—the so-called critical periods of development (Widdowson and McCance 1975). Thus, providing testosterone to female birds during a critical period of her life will make her develop a capacity for singing, an otherwise male feature. Exposure to deleterious environmental compounds, or child abuse during such critical periods, can induce specific developmental alterations, and disease susceptibility to the offspring, and such disease susceptibility can be transmitted to subsequent generations. It will be later shown (in Chapter 10, Section B) that there are critical periods during embryogenesis and early postnatal life when epigenetic processes are susceptible to perturbations by maternal nutrition. I will claim in this chapter that transitions between life-history stages are critical adaptive periods of life (Hochberg 2009).

Defining transitions between life-history stages occupies a place somewhere between myth and reality in child growth and development. Mostly, the onset of adolescence has been described and discussed as if it starts abruptly, the so-called onset of puberty. It is easy to show that turning points are in fact deliberate constructions that exaggerate some aspects at the expense of others and that focusing on onsets of life stages usually obscures a great deal of continuity: *natura non fecit saltus*—nature does not leap—has been a principle of natural philosophy since ancient times. It appears as an axiom in the works of Gottfried Leibniz (1646–1716) and Isaac Newton (1643–1727), as they developed infinitesimal calculus. It is also an essential element of Darwin's treatment of natural selection in his *Origin of Species*. Rather, nature makes gradual transitions.

And yet, staging has retained a central place in evolutionary life-history theory and in this endocrine deliberation of life-history theory, and rightly so. In defining a life-history stage for evolutionary, physiological, social, or endocrine inquiries, one inevitably needs to define the stage's beginning and end. Transitions are a key concept of that narrative, whereas offsets of a developmental stage remain even more intricately difficult to characterize. In fact, offsets of life-history stages may not really exist in normal physiology, other than in response to environmental cues or therapeutic measures, as in the treatment of children with precocious puberty.

Multiple hormone mechanisms have evolved to activate heterochrony, both physiological and behavioral traits, at the right time and in the correct context as the child transits from infancy to childhood, then to juvenility, and later to adolescence and youth. With detailed investigation of organisms' life histories in their natural habitat, one can determine the potential ecological costs underlying hormone–physiology–behavior interactions at stage transitions that, in turn, shed light on their evolution (Wingfield, Jacobs et al. 1997).

Analysis of growth through derivation of observed measurements allows for plotting the growth velocity chart (first derivation of the growth chart; Fig. 2.B.1)

(CDC 2000). The stages of child growth become evident, with the rapid deceleration of infantile growth, stable growth during childhood, decelerating growth during juvenility and acceleration of growth during adolescence, when it reaches the peak growth velocity, before deceleration to cease growing. Life history may also be revealed from the germ cell count in the testes. Figure 2.B.2 shows the clear rebound at childhood and adolescence onset.

As the child transits from infancy to childhood, the growth hormone–insulin-like growth factor-I axis sets in and new behavioral traits develop, but the metabolic forces that drive infantile growth remain as active and the infantile mental achievements develop further, though some infantile manners recede while the child ceases to feed on mother's milk. At the transition from childhood to juvenility, while the juvenile joins adults in their daily activities, adrenal androgens come into play for the first time—the so-called adrenarche—and behaviors change, but the growth hormone–insulin-like growth factor-I axis remains as imperative. Transition to adolescence is associated with the onset of the hypothalamic–pituitary–gonadal axis activity to induce puberty, while both adrenal androgens and the growth hormone–insulin-like growth factor-I axis continue to play a role in growth.

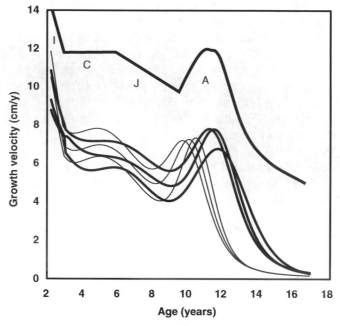

Fig. 2.B.1. The first derivative of human height–growth velocity charts. The 3rd, 50th, and 97th percentile of first derivatives for both boys (thick lines) and girls (thin lines) were calculated from the U.S. CDC 2000 data (CDC 2000). The upper part indicates the four youth life-history stages: infancy (I), with its decelerating growth velocity; the stable growth of childhood (C); the decelerating growth of juvenility and the ups and down of growth velocity during adolescence (A).

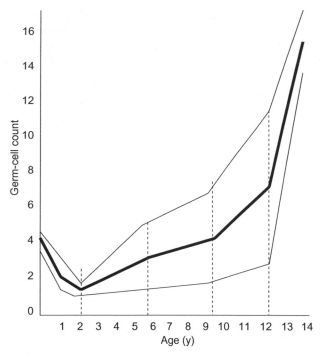

Fig. 2.B.2. Life history stages as revealed from germ cell counts in human testes. Data from Huff, Hadziselimovic et al. (1987).

Endocrine mechanisms are available to set off any of these means in adjusting to environmental signals: malnutrition and psychosocial deprivation will slow the growth hormone–insulin-like growth factor-I axis activity, anorexia nervosa will block adrenal and hypothalamic–pituitary–gonadal androgen secretion, and physical exertion will slow the hypothalamic–pituitary–gonadal axis, to name just three possibilities. (For more on the endocrinology of each of these stages, see the relevant chapters).

C. DEVELOPMENTAL PLASTICITY AND ADAPTATION

Observe always that everything is the result of a change, and get used to thinking that there is nothing Nature loves so well as to change existing forms and to make new ones like them.

—Marcus Aurelius (Meditations, iv. 36)

Every living organism has two histories that determine its biology: an evolutionary history whose duration is in the hundreds of thousands of years, and a developmental history that begins with conception. Evolutionary changes are slow, whereas environmental changes can be quick. In fact, when facing the challenge of developing an individual that best fits its environment, nature demonstrates an interesting combination of five different adaptive processes that influence human phenotype, and each process operates on a different time scale (Muller 2007; Hochberg, Feil et al. 2011). The first adaptive process involves changes in gene sequence and frequency in a population or species, and this process of adaptive genotype occurs over several hundred thousand years. Suppose a certain gene mutation conferred a selective advantage of 5%; how long would it take to increase from a frequency of 1% to 90%? The answer is 325 generations or 8,000 years (Crow 1961). For this type of adaptation, time is an important constraint. The second process is the modification of homozygosity of the population, and this process occurs over several hundred years and numerous generations. The third process refers to plastic/adaptive phenotype, and this process occurs over the total life span of the individual, and may be carried forward for three to four generations. Yet, our evolved state is often mismatched to our modern environment because that environment is changing more rapidly than these three adaptive mechanisms. Finally, the fourth process is short-term acclimatization that can last several months or years. On an even shorter scale, the developmental history of an individual organism is associated with new structures that appear that cannot be explained in terms of the unfolding or growth of structures, which are already present in the egg at the beginning of development, programming the organism for later ecologically appropriate physiology. Fifth in the list of adaptive mechanisms is cultural adaptation, which will be discussed in the next chapter.

As genes are modified in their sequence during evolution, they change the structure, amount, or timing of protein generation. These proteins will cause one result in environment A, and they may cause a different result in environment B. These different results define plasticity, which includes adaptive accommodation in all aspects of the phenotype, including morphology as well as physiology and behavior. Plasticity in developmental programming has evolved in order to provide the best chances of survival and reproductive success to the organism in an ever-changing environment. The role of flexibility in facilitating evolutionary changes has been noted by the American philosopher and psychologist James Mark Baldwin (1861–1934), who called it "organic selection" or "functional selection," also known as the Baldwin effect, deriving from the models inspired by "divine pre-establishment" of the Dutch Jewish philosopher Baruch Spinoza (1632–1656). Originally linked to the philosophy of mind, Baldwin suggested in his *Mental Development in the Child and the Race* that selected offspring would tend to have an increased capacity for learning new skills rather than being confined to genetically coded, relatively fixed abilities. It placed emphasis on the fact that the sustained behavior of a species or group can shape the evolution of that species. In more general terms, it implied an evolutionary fitness enhancement due to phenotypic accommodation. Over the years this was called "regulative ability," "phenotypic compensation," "ontogenetic buffering," and "evolvability." But whatever the name, phenotypic accommodation is the adaptive plasticity in adjusting variable aspects of the phenotype, without change of the gene sequence (West-Eberhard 2005). This definition is much in line with the concept of epigenetics, which will be discussed later. Phenotypic accommodation can facilitate the evolution of novel morphology by alleviating the negative effects of change, and by giving a head start to adaptive evolution in a new direction. It is the result of adaptive developmental responses, so that the novel morphologies that result are not random variants, but to some degree reflect past functionality (West-Eberhard 2005).

The *Homo sapiens sapiens* has been quite unique among organisms as a single species that lives under a variety of environmental conditions unprecedented in nature, including the entire range of geographical latitudes and altitudes, as well as extremely diverse weather conditions. Whereas the above vary slowly over generations, nutritional conditions may change more rapidly, and evolution has provided for mechanisms to adapt to these extremes, while sociocultural adjustments filled the remaining gaps when the environment changes faster than the evolutionary time scale.

The secular trend in child growth and puberty is a dazzling example of such an adaptation (Figs. 2.C.1 and 2.C.2). Over the six generations of a century and a half between the mid-1850s and the year 2000, genes have obviously changed little, and the secular trend exemplifies the degree of available plasticity. If Italian men are now 13 cm taller than they were 150 years ago, and if they indeed reached a saturation point in their height, the plasticity magnitude for height is at least 13 cm. The 4-year difference in first menstruation age (menarche) between 19th- and 20th-century girls demonstrates the plasticity magnitude for this event with a given genome, and the declining age of menarche does not seem to have reached a saturation point.

Adaptation of height to the environment is also evident in skeletons of ancient Maya that were found in Tikal, Guatemala (Haviland 1967). Tomb skeletons (indicating wealth) from the classical period 250–400 AD show an average male length of 172 cm, while the poor non-tomb skeletons are 164 cm long. In the late classical period 700–900 AD, tomb skeletons are on average 162 cm and non-tomb skeletons

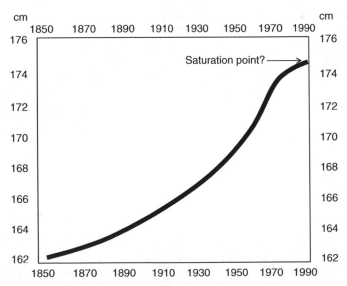

Fig. 2.C.1. The secular trend in height. Stature at the age of 20 years in Italy from 1854 to 1990, showing the secular trend in height. Height seems to have almost reached a saturation point in 1990. Data from Arcaleni (2006).

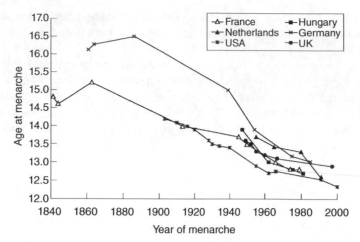

Fig. 2.C.2. The secular trend in puberty. Declining age of menarche in some European countries and the US from 1840 to 2000. Adapted with permission from Bellis, Downing et al. (2006). The lines do not show a saturation point; the trend is expected to continue.

are 156 cm, showing the impact of nutrition on growth of the rich Maya as compared to the poor. Post conquest (1520), the average length of skeletons is 156 cm; by then everyone became poor.

A given genotype can generate multiple phenotypes (polyphenism[5]) depending on the environmental conditions experienced by the organism during development. Such phenotypic plasticity has been described in almost every group of plant and animal, including vertebrates and humans (Crespi and Denver 2005). Adaptive phenotypic plasticity allows individuals within a population to accommodate a changing environment by facilitating rapid movement to a new fitness optimum (Price, Qvarnstrom et al. 2003). In the most extreme cases, plastic changes of a trait are able to completely move a population to a new fitness optimum, and no genetic adaptation is required. In cases where a plastic change in the mean is not sufficient to shift a population to a new optimum, it can allow a population to persist until sufficient adaptive genetic changes occur (Yeh and Price 2004). Plastic responses during early development have particularly important fitness consequences since they can lead to permanent and profound modifications of morphology or physiology (Nijhout 2003). In fact, for most trade-offs to function, they must be delineated early in development. The impacts of nutritional cues on phenotypic development and the expression of traits such as obesity, metabolic syndrome, diabetes mellitus type 2, and reproductive performance have been extensively investigated (Gluckman, Hanson et al. 2008). Likewise, it has been claimed that there is a relationship between nutritional status in late childhood and the age of menarche.

The German-American anthropologist Franz Boas (1858–1942) introduced the evolutionary concept of developmental plasticity early in the 1900s as part of his study of human cranial structure; this was expand in the 1960s by Lasker (1969). We

---

[5] Polyphenism: Evolved adaptations in which a genome produces discrete alternative phenotypes in different environments, such as in metamorphosis.

now recognize two classes of adaptive responses or plasticity (Gluckman and Hanson 2007a; Hochberg, Feil et al. 2011). The first class is anticipatory or predictive adaptive responses, when the developing organism uses environmental cues to forecast the future environment in order to adjust its phenotypic trajectory accordingly. The second class is immediate adaptive responses, which promote short-term maternal or fetal survival, but at a cost of trade-offs later in life (developmental plasticity). The window of developmental plasticity extends from before conception to weaning, and involves epigenetic responses to environmental influences that exert their effect in early life. These two adaptive responses do not come without a significant cost. Within the potentially adaptive class of responses, the organism may engage in a trade-off between phenotypic changes in order to ensure short-term survival, but at the expense of a long-term advantage. Therefore, trade-offs may manifest themselves with longevity at the cost of reduced survival of the juveniles. Trade-offs arise because energy has to be parcellated, and individual members of a species make a virtual cost-benefit analysis in order to determine the true value of an adaptive response. Such is the consequence of embryonic fetal development that occurs in a deprived intrauterine environment as a result of limited transplacental nutrient supply. In response, the fetus protects the development of the heart and brain, but at the expense of other organs, resulting in somatic growth retardation. Underlying this developmental plasticity is the fundamental premise that the physiology of an individual is driven by the induction of a particular developmental program, which is influenced by the prevailing environment that exists during a critical developmental period (Bateson, Barker et al. 2004). While more likely to survive to birth, such neonates are smaller and have higher rates of morbidity and mortality. Given the long-term trade-off in fitness, intrauterine growth restriction (IUGR) is an example of when the immediate adaptive response to environment is cryptically maladaptive (Gluckman and Hanson 2007b). Under severe situations in polytocous species, their adaptation may be driven by maternal interests rather than offspring interests (Haig 2010), the so-called parent-offspring conflict, but this seems unlikely in slow reproducing monotocous species (Gluckman and Hanson 2007a, 2007b; Gluckman, Hanson et al. 2008).

When the environmental cue is subtle in early life, there is frequently no immediate adaptive response, because the cue is interpreted as a surrogate predictor of the later reproductive environment. As developmental plasticity is limited by temporal constraints, this interpretation of the response can create a situation where there is evolutionary advantage for the fetus to adjust its phenotypic development in order to develop a "better-matched" postnatal phenotype. While such processes are selected robustly across taxa, there is a higher risk of lower fidelity to the prediction in mammals. In fitness terms, this need may not have much impact because health and fitness are distinct concepts.

However, for humans who live beyond the peak reproduction, health consequences, rather than reproductive consequences, become the primary concern. For example, when the nutritional status in adulthood changes markedly from that experienced during development, and the expectation of a poor environment is not subsequently met, metabolic disease is more likely (Barker 2006). Life-history theory would argue that energetic- and stress-related cues, such as predators, that are experienced by the mother are likely to be primary environmental triggers of developmental plastic responses. Predictive responses are induced primarily by subtle cues, and immediate adaptive responses are induced by more severe cues. However, it is to be anticipated that both types would coexist if the latter are induced. I have previously mentioned the short-statured hunter-gatherers Ju!'hoansi San of the Kalahari. Their investigator, Nancy Howell, suggested that "the small body size may be a positive consequence of a culture-wide adoption of a program of caloric restriction, which has been shown to increase health and longevity in modern human populations, and which is the only reliable method of reducing morbidity due to chronic disease and thus extending longevity into old age known to work in industrialized societies" (Howell 2010). She suggested that their cultural practices concerning food and food preparation discourage eating beyond the amount needed in any given day was relevant, and that this was true for both adults and children. Ju!'hoansi San mothers did not encourage them to eat beyond the minimum needed to satisfy their appetite and did not prepare particularly delectable meals.

As a consequence of life conditions under changing environments, children may be stunted for short or longer periods, be underweight or overweight, and be at risk for disease. In the endocrine jargon, this has been labeled "developmental programming." The evolutionary language for the same is a "predictive adaptive response" (Gluckman and Hanson 2005). Even under good conditions, the stages of human life history are replete with trade-offs for survival, productivity, and reproduction. Under adverse conditions, trade-offs result in reduced survival, poor growth, constraints on physical activity, and poor reproductive outcomes.

Whereas programming implies permanent maladaptive effects that place people at risk for disease, or a pathology in medical jargon, predictive adaptation considers the phenotypic changes as constructively adaptive, and at two levels: (1) short-term adaptive responses for immediate survival, and (2) predictive responses required to ensure postnatal survival to reproductive age to increase reproductive fitness. To adapt to a changing environment, an organism has to be able to modify the expression of its genome in a stable manner. It is the changeability that is genetically determined whereas the changes determine the phenotypic plasticity. Organismal size, including child growth, is probably the supreme case in point of this paradigm.

The initial (Ravelli, Stein et al. 1976) and then the plethora of reports on the consequences of the 1944–1945 Dutch famine during World War II illustrate these concepts at their extreme. Exposure to a relatively short famine of 6 months early in life had lifelong effects on health, and these effects varied depending on the timing and length of exposure and its recovery timing and period. Exposure to famine during gestation resulted in later impaired glucose tolerance, obesity, coronary heart disease, hyperlipidemia, and hypertension, but also in antisocial personality, schizophrenia, and affective disorders (Stein, Pierik et al. 2009). Exposure to famine during childhood resulted in changes in reproductive function, earlier menopause, and changes in insulin-like growth factor-I levels.

Following the initial description by Barker, Osmond et al. (1989), another plethora of reports described the health consequences of intrauterine growth retardation. Intrauterine programming became the doctrine that explained these phenomena: The ultimate outcome is an evidence of pathology. Well-nourished mothers have offspring who are adapted to affluent conditions, whereas mothers on a low level of nutrition have offspring who are adapted to lean environments (Bateson 2007). If the mother's "forecast" of her offspring's future environment is incorrect, the health of her offspring in a mismatched environment may suffer severely (Gluckman and Hanson 2007b). Understanding the evolutionary background places the developmental origins of ill health in humans in context and has profound implications for public health (Bateson 2007). This concept is further discussed in Chapter 3, Section C.

When a population encounters a new environment, plastic responses alter the distribution of phenotypes, and hence affect both the direction and the intensity of selection. Plastic responses in one trait result in novel selection pressures on other traits, and this can lead to evolution in completely different directions than those predicted in the absence of adaptive plasticity. Kuzawa (2007) proposed that an organism has metabolic potential in excess of survival requirements that he called "productivity," supporting growth before being shunted into reproduction after growth ceases. In his intergenerational phenotypic inertia model, he predicted that plasticity in growth rate will be positively correlated with components of future adult reproductive expenditure. Thus, early nutrition or growth rate predicts offspring size in females, and increased somatic investment related to reproductive strategy in males. Indeed, population birth weight and sexual size dimorphism are predicted to increase in response to improvements in early nutrition (Kuzawa 2007). This is perpetuated not merely during the life cycle but across generations: In females, growth rate predicts future nutritional investment in reproduction, which in turn determines fetal growth rate in the next generation. Kuzawa suggested that growth and reproduction serve as mutually defining templates, thus creating a phenotypic bridge, and allowing ecologic information to be maintained during ontogeny and transmitted to offspring (Kuzawa 2007). Resetting of metabolic production in response to maternal nutritional cues may serve a broader goal of integrating transgenerational nutritional for adjusting long-term strategy.

Yet, the information passing to the offspring does not represent accurate cues of external environmental quality but rather comprises a summary of maternal phenotype that reflects the mother's own development in buffering her fetus (Wells 2003). Known as Wells's "maternal fitness approach," this strategy is ideal for a species in which the offspring remains dependent on maternal energy allocation for many years following birth. In humans, this period is restricted to infancy, as defined by weaning from breast-feeding.

Investigation of stock size in the Dungeness crab (*Cancer magister*) is a classic example of phenotypic accommodation (Shanks and Roegner 2007). During a 5-year sampling period, adult crab population size from Oregon through central California varied directly with the number of terminal-stage larvae returning to Coos Bay in Oregon 4 years earlier. More than 90% of the variation in the number of returning crabs was explained by the timing of the spring transition, a seasonal shift in atmospheric forcing that drives ocean currents along the California coast. Early spring transitions led to larger numbers of returning Dungeness crabs, while in four other crab species with very different life-history characteristics, early-spring transitions led to lower numbers of returning larvae. The size of the commercial catch of Dungeness crabs is significantly and negatively correlated with the date of the spring transition throughout the California Current System.

The timing of metamorphosis, hatching, or birth depends on the trade-offs between growth opportunity and mortality risk in the developmental habitat. Metamorphosis of the amphibian is a paradigm of polyphenism of environmental effects on a period of transition and its adaptive plasticity. It was shown that population density, food availability, habitat desiccation, and exposure to predators have significant effects on body size and the timing of metamorphosis (Crespi and Denver 2005). If conditions are favorable in the larval habitat, tadpoles continue to capitalize on growth opportunities and delay metamorphosis, whereas accelerated metamorphosis translates into smaller body size and lower survival rates. The environmental stimulus alters the endocrine mechanism of metamorphosis by altering either the pattern of hormone secretion or the pattern of hormone sensitivity in different tissues. Such changes in the patterns of endocrine interactions result in the execution of alternative developmental pathways (Nijhout 2003).

The great German-American evolutionary biologist Ernst Mayr (1904–2005) referred to phenotypic accommodation to the environment as "soft inheritance," in contrast to "hard inheritance" by gene structure. However, hard evidence is now available for molecular mechanisms that underlie adaptive plasticity, which is referred to in Chapter 11, Section B, on epigenetics.

In that respect, it is interesting to mention an illustrative example of such developmental plasticity: the appearance of the "helmeted" morph in the freshwater crustacean *Daphnia* in response to increased presence of predators (Latta, Bakelar et al. 2007). This study showed a reduction in egg number and body size of the *Daphnia* in response to a chemical cue—coming from an introduced predator. These phenotypic changes have a genetic basis but are partly due to a direct response to chemical cues from fish via adaptive phenotypic plasticity. Body size showed the largest phenotypic change, on the order of nine phenotypic standard deviations, with approximately 11% of the change explained by adaptive plasticity. Both evolutionary and plastic changes in body size and egg number occurred, but no changes in the timing of reproduction were observed.

In mammals, such adaptive plasticity is typified by the fetal meadow vole (*Microtus pennsylvanicus*), which determines the appropriate thickness of its postnatal coat *in utero* in response to maternally derived signals of day length, a predictable indicator of seasonal change (Lee, Lawler et al. 1987). The recognition that environmental cues may profoundly influence the development and appearance of functional morphs in the population encourages a "post-genomic" interpretation of phenotype. Specifically, it challenges the simplistic interpretation of phenotype as a deterministic fixed outcome of the genotype, which has dominated much of developmental and evolutionary biology thinking in the 20th century. Phenotype viewed rather as "the expression of a given genotype under its particular environmental influences" (Sultan and Hasnain 2003) not only reignites the arguments of earlier evolutionary biologists such as Ivan Ivanovich Schmalhausen (1884–1963) and Conrad Hal Waddington (1905–1975), but also sits comfortably with emerging notions in modern molecular biology. The almost exponential expansion in our understanding of epigenetic gene regulation is providing potential mechanistic insight to phenomena such as genotype/environment interactions and developmental plasticity.

A discussion of phenotypic plasticity invariably brings up the issue of stem cells and adaptive plasticity, which remains a contentious issue in biology today. The term "plasticity" here refers to the capacity of tissue-derived stem cells to regenerate and reprogram—a phenotypic potential that extends beyond the differentiated cell phenotypes of their resident tissue (Bornstein 2006). Thus, stromal stem cells possess a plasticity to mature into adipocytes, chondrocytes, or osteocytes. It was suggested that this pluripotency reflects a fundamental characteristic of cellular diversity, which is manifested as the adaptive response to a functional pressure by the cell's biochemical and biophysical microenvironments that would drive their differentiation (Moldovan 2005). In this model, differentiation is a dynamic, reversible, and open-ended process where the cells would maintain the flexibility to respond to changing environmental cues with a fine-tuning of their structure, a property that was previously called "cellular plasticity." However, they also possess the potential to change directions altogether and differentiate into very different cells, such as pancreatic beta-cells differentiating in a liver. Thus, differentiation of stem cells may be a form of cellular plasticity within the larger context of adaptive plasticity, whereas their stemness remains associated with self-renewal (Moldovan 2005). For more on this topic, see below, Section F.

D. CULTURAL ADAPTATION TO THE ENVIRONMENT

Human cultures and technologies have modified life. A central principle of anthropology is that relations between humans and the environment are mediated by culture: the so-called cultural ecology. Even in plain foraging societies, members cooperate in large groups by division of labor, care for the sick and disabled, and a moral code usually shared and enforced. Humans learn from one another in a wide variety of domains. Consequently, information and behaviors are culturally transmitted in human populations. For scientists, the concept is straightforward. Each generation of scientists builds on the achievements of their predecessors in a regular, ascent of complexity.

Like most other aspects of evolutionary biology, Charles Darwin saw it clearly. In *The Descent of Man*, he drew direct parallels between biological evolution and the cultural evolution:

> ... although a high standard of morality gives but a slight or no advantage to each individual man and his children over the other men of the same tribe, yet that an increase in the number of well-endowed men and an advancement in the standard of morality will certainly give an immense advantage to one tribe over another. A tribe including many members who, from possessing in a high degree the spirit of patriotism, fidelity, obedience, courage, and sympathy were always ready to aid one another, and to sacrifice themselves for the common good, would be victorious over most other tribes; and this would be natural selection.

Cultural adaptation is much more rapid than genetic adaptation. Adaptive cultural processes are strongly evident in the initial hundreds of years after migration, until modifications in population homozygocity become evident. As a general role, the more recently a group has migrated into an area, the more extensive its cultural, as opposed to biological, adaptation to the area will be.

For example, when the light-skinned Arabs arrived in the Arabian Peninsula from the north some 2,500 years ago, they had to adapt to a heavy radiation impact. On the other side of the Red Sea, Sudanese had dark skin to protect them from ultraviolet radiation. Such radiation is known to degrade folic acid, an essential factor for cell proliferation. Folic acid deficiency is known to arrest spermatogenesis, with obvious impact on reproductive fitness. To culturally adapt to intense UV exposure the newly arrived light-skinned migrants started using heavy protective clothing and portable shades (tents). They mostly had to protect women, who have lighter skin color around the globe, to enable them to make more vitamin D for greater lifetime calcium need in pregnancy and lactation. Thus, women of the Peninsula were and still are confined to their shaded houses and heavily covered when leaving the house.

Many of the diseases associated with contemporary Western populations, and spreading across the globe, have arisen through mismatch between our ancient genetically influenced biology and the cultural, dietary, and physical activity patterns of modern societies. The culture in agriculture has certainly made an immense impact on human evolution. The global epidemic of obesity represents a combination of rapidly changing, culture-based, behavioral changes on the background of discordant genomic predispositions. The Framingham Heart Study showed how obesity spreads and is clustered among relatives and friends (Christakis and Fowler 2007). The study showed that if one friend became obese, the chances his friends would become obese increased by 171%. The risk that the friend of a friend of an obese person would be obese was 20% higher than in a random network. If one sibling became obese, the chance that the other would become obese increased by 40%, and if one spouse became obese, the likelihood that the other spouse would become obese increased by 37%.

Cultural adaptation must ultimately be rooted in our biology: It has been attributed to uniquely human social learning mechanisms (Whiten and Mesoudi 2008). Thus, culture provides a second inheritance system for human behaviors. The relationship between environment, genes and culture is rather complex and to a great extent still obscure.

E. ADAPTIVE PLASTICITY OF ATTACHMENT BEHAVIORS

An important environmental cue for infants and young children is the caregiving behaviors of their parents, which can be used as a predictive indicator of the security of their environment (see, e.g., John Bowlby's *Attachment and Loss*, 1969–1982). The resultant attachment patterns are transmitted transgenerationally (Belsky and Fearon 2002; Del Giudice 2009). Such cues are subject to parent–offspring conflicts and genomic imprinting on human neurodevelopment and behavior that are central to evolved systems of mother–child attachment (Haig 2010). The degree of security that is experienced during childhood sets development on alternative pathways and adaptively shapes the individual's future reproductive strategy. A secure attachment will result in a reproductive strategy that is based on late maturation, a commitment to a long-term relationship, and a large investment in parenting. In terms of evolutionary developmental biology (evo-devo), which studies the developmental mechanisms that control body shape and forms and the alterations in gene expression and function that lead to changes in body shape and pattern, the expected response to a secure environment will include investment in large body size (Goodman and Coughlin 2000). This example of transgenerational phenotypic plasticity contrasts that of an insecure attachment and a small parental investment that involves a large number of children: The response is a compromise in body size, early reproduction, and short-term mating.

Attachment is a two-hit system, with important adaptations during infancy and juvenility (Del Giudice 2009) (Fig. 2.E.1). In stressful conditions, parenting style becomes harsher and less sensitive and marital discord increases, causing the child

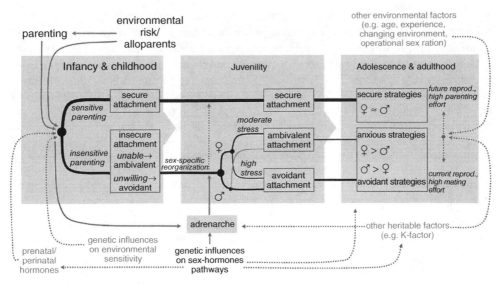

Fig. 2.E.1. Attachment-related reproductive strategies. Schematic diagram of the development of reproductive strategies based on the two-hit model of parental behaviors during infancy and juvenility. Black lines represent typical developmental pathways, with thickness roughly proportional to phenotype frequency in relatively low-risk populations. Adapted with permission from Del Giudice (2009).

to experience chronic psychosocial stress that would lead to insecure attachment patterns (Belsky, Steinberg et al. 1991). Insecure children thus receive such indirect information about their milieu: that resources are in short supply and erratic, that people cannot be trusted, and that mating relationships tend to be short and uncommitted. This switches development toward a reproductive style based on opportunistic interpersonal orientation, early reproduction, and low parental investment (offspring quantity vs. quality). Secure attachment/low stress, on the other hand, leads to delayed mating, high parental investment, and a trusting and reciprocally oriented attitude.

The attachment system is a characteristic of a societal pattern of behaviors, but there is much individual variation and sexual dimorphism in the organization of actual attachment relationships. This may be a result of asymmetries in parental investment and sexual choices (Belsky and Fearon 2002). Sex differences in attachment arise in juvenility, starting from about 6–7 years of age (Del Giudice 2009). During this stage of sex-biased reorganization of the attachment system, a majority of insecure girls shift to ambivalent attachment and most insecure boys shifting to avoidant attachment.

F. NOTE BY GEORGE CHROUSOS ON STRESS IN EARLY LIFE: A DEVELOPMENTAL AND EVOLUTIONARY PERSPECTIVE

Early life, including the fetal, childhood, and puberty/adolescence periods, is characterized by marked changes in growth and development. The former includes physical changes in stature and body weight, shape and composition, as well as the process of puberty, and is usually completed by the age of 14–16 years. The latter concerns the psychosocial, intellectual, and emotional maturation of the individual, including completion of brain development, and is finished by the age of 24–27 years. "Stress" defined as the "state of threatened or perceived threatened homeostasis" may influence growth and development, producing both concurrent and/or future psychopathology. To understand the profound influences that stress in early life has on the individual throughout his or her life, it is crucial to briefly review the concepts of stress and homeostasis, explain the mechanisms through which the damage is done, and finally concentrate on the somatic and behavioral pathology caused.

1. Stress Concepts

Humans, as highly complex living systems, are constantly in a state of an extremely complicated equilibrium, which in the early 20th century, Walter Cannon called "homeostasis," i.e., "steady state," a term that supplanted the ancient Greek term "isonomia," i.e., "equal division" (Chrousos, Loriaux et al. 1988; Chrousos 2009) (Fig. 2.F.1). As long as we live, homeostasis is maintained; its complete dissolution means death. Homeostasis is frequently threatened by disturbing forces, the "stressors," which can be physical or emotional. The organism has its own means, the "adaptive response," to counteract these forces and reestablish the steady state.

Up to a certain threshold of a stressor strength and duration, the adaptive response is able to bring the usual healthy homeostasis or "eustasis" back from a disturbed state (stress or "distress") completely, without any cost to the individual (Chrousos 2009) (Fig. 2.F.2). On the other hand, when a stressor cannot be entirely counteracted by the adaptive response of the individual and the homeostasis attained is suboptimal, and, in fact, associated with harm to the individual, this state is called

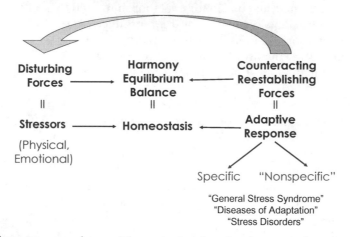

Fig. 2.F.1. Direct correspondence of the ancient and modern homeostasis and stress concepts. Now when we say "stress" we mean a "state of threatened or perceived as threatened homeostasis." The specificity of the adaptive response is lost as the magnitude of the stressor increases. An excessive or inadequate adaptive response may itself be a stressor disturbing homeostasis (top arrow). The concepts of "general stress syndrome,, "diseases of adaptation," or "stress disorders" were proposed by H. Selye. From Chrousos, Loriaux et al. 1988, modified.

Homeostasis over Time

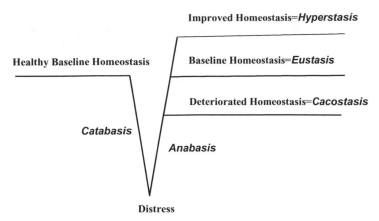

Fig. 2.F.2. The three adaptive responses to stress. After exposure to a stressor and descent (*catabasis*) into disturbed homeostasis (*stress* or *distress*), a healthy organism's adaptive response brings homeostasis back (*anabasis*), with one of three possible outcomes: (1) the adaptive response succeeds and returns the organism to its previous healthy homeostasis (*eustasis*); (2) the adaptive response succeeds partly, either because it is inadequate or because it is excessive and/or prolonged, and the organism returns to an unhealthy form of homeostasis (*cacostasis*) that is damaging the host, producing frailty and curtailing life expectancy; and (3) the adaptive response succeeds and the organism gains from the experience returning to an improved homeostasis, a superior state that is to the benefit of the host, improving its resilience and chances to survive (*hyperstasis*).

"allostasis" (different state), or, more accurately, "cacostasis." Finally, the human organism may learn from a stressor, adapt and frequently attain and sustain a homeostasis that is better than would have been expected by its genetic and epigenetic constitution and the salient environment, i.e., a superior state called "hyperstasis."

There is a wide spectrum of human stressors ranging from the daily hassles to life transitions, to starvation and bereavement (Chrousos 2009) (Table 2.2). For the developing human one should consider all intrauterine stressors, lack or loss of caregiver support, defective attachment with the caregiver, neglect, physical and sexual abuse, and all kinds of psychosocial trauma, including parental divorce, poverty, substance abuse, etc.

2. Stress Mechanisms

In the case of the human organism, as in all mammals, the adaptive response is subserved by a specialized homeostatic system in our brain and body—the "stress system." which is activated to help us deal with stress when a stressor of any kind exceeds a certain threshold (Chrousos 2009; Chrousos and Gold 1992) (Fig. 2.F.3). The principal central nervous system (CNS) effectors of the stress system, which are highly interlinked, include the hypothalamic corticotropin-releasing hormone (CRH), arginine vasopressin (AVP), and proopiomelanocortin-derived peptides (alpha-melanocyte-stimulating hormone and beta-endorphin), and the brainstem *locus caeruleus*-norepinephrine/autonomic nervous system (LC/NE) centers.

TABLE 2.2. Human Stressors

- Daily hassles
- Natural and unnatural catastrophes
- Life transitions (fetal life, early life, puberty, menstrual cycle, pregnancy and postpartum, menopause)
- Lack or loss of caregiver support in early life (attachment, bonding, neglect, abuse)
- Starvation, excessive exercise, excessive nutrition, obesity
- Low socioeconomic status
- Job loss, downsizing, sense of loss of control
- Bereavement/taking care of sick relatives
- Addictions/toxic substances
- Inflammations (traumatic, infectious, autoimmune, allergic)
- Anxiety, depression, personality disorder

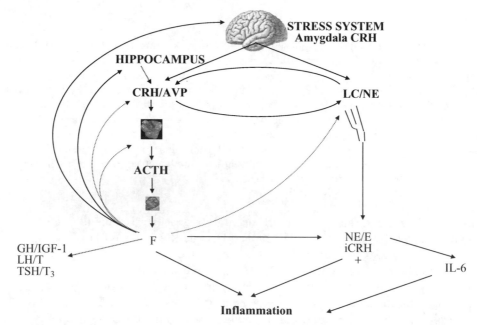

Fig. 2.F.3. CNS, inflammatory, and endocrine stress connections. The stress system, along with its central and peripheral components and its mediators is generically activated when homeostasis is threatened or perceived as threatened and helps the organism to return to homeostasis. Its optimal response should be proportional to the stressor and time limited. Interestingly, corticotropin-releasing hormone (CRH) plays a major role in the amygdala stimulating fear/anger and the periphery as a proinflammatory mediator degranulating mast cells. Interleukin-6 (IL-6), a traditional inflammatory cytokine and a major mediator of the "sickness syndrome," is also activated in stress and contributes to its manifestations. For simplicity purposes the parasympathetic limb of the autonomic nervous system has not been included. During stress, this system may withdraw its activity, in effect assisting the sympathetic system, or increase its activity acting as a break. LC/NE = *locus caeruleus*/norepinephrine-sympathetic system; CRH = corticotropin-releasing hormone; AVP = arginine-vasopressin; ACTH = corticotropin; F = glucocorticoid, cortisol; NE = norepinephrine; E = epinephrine; iCRH = immune CRH; IL-6 = interleukin-6; GH = growth hormone; IGF-1 = insulin-like growth factor-I; LH = luteinizing hormone; T = testosterone; FSH = follicle-stimulating hormone; E2 = estradiol; P4 = progesterone; TSH = thyrotropin T3 = triodothyronine. From Chrousos 2000; Chrousos 2009 (modified).

The principal peripheral effectors of the adaptive response to stress are the end-hormones of the hypothalamic–pituitary–adrenal (HPA) axis, glucocorticoids, and of the systemic and adrenomedullary sympathetic nervous systems, the catecholamines norepinephrine and epinephrine (Chrousos 2009; Chrousos and Gold 1992). Interestingly, postganglionic sympathetic nervous system fibers also secrete, among other substances, peripheral (immune) CRH, while both catecholamines stimulate interleukin (IL)-6 release by immune and other peripheral cells via beta-adrenergic receptors.

The target tissues of all these stress mediators are in the brain and periphery, including all other homeostatic systems, which generally operate within an optimum range, i.e., their own complex equilibrium, with too little or too much activity representing cacostatic states. Indeed, these systems malfunction and lead to suboptimal adaptation at both sides of an inverse U curve (Chrousos, Loriaux et al. 1988; Chrousos 2009; Chrousos and Gold 1992) (Table 2.3). Thus, the homeostasis of executive/cognitive, fear/anger and reward systems and the wake/sleep centers, the growth, thyroid and reproductive axes, as well as the gastrointestinal, cardiorespiratory, metabolic, and immune systems are markedly affected by the stress system (Chrousos 2009; Chrousos and Gold 1992).

Generally, the adaptive response to stress is acute or temporally limited and meant to help with a stressor, returning to its standard circadian resting state as soon as possible (Chrousos and Gold 1992). CRH administered intracerebroventricularly to several experimental animal species, reproduced the phenomenology of the adaptive response to stress as summarized in Table 2.4. In a series of subsequent studies, it also became apparent that the hypothalamic CRH/AVP and brainstem LC/NE centers of the stress system mutually innervated and stimulated each other with CRH and AVP or norepinephrine, respectively (Fig. 2.F.3). This positive, reverberating feedback system could, therefore, be activated with CRH, norepinephrine, or any other stimulus that could set into motion either side of this highly complex but integrated brain loop.

TABLE 2.3. Adaptive Homeostatic Systems can Malfunction and Lead to Suboptimal Adaptation on Both Sides of an Inverse U Curve

Low = Cacostasis	HOMEOSTATIC SYSTEM	High = Cacostasis
Hypoarousal	**Stress-CRH/LC-NE**	Hyperarousal
Executive/Cognitive Dysfunction	Executive/Cognitive-F/PFL,	Executive/Cognitive Dysfunction
No Behavioral Inhibition	Amygdala-Fear/Anger	Behavioral Inhibition
Depression	MCLS-Reward	Depression
Hypotension	Cardiorespiratory	Hypertension
Insulin Resistance	Metabolic	Insulin Resistance
TH2 Auto-immune disease, TH1 Infectious disease	Immune	TH1 Auto-immune disease, TH2 Infectious disease
Fatigue, Hyperalgesia	Fatigue-Pain	Fatigue, Hyperalgesia
Sleep Disturbances (Somnolence)	Sleep	Sleep Disturbances (Insomnia, Awakenings)

LC/NE = *Locus caeruleus*/norepinephrine–sympathetic systems
F/PFL = Frontal/prefrontal lobes
MCLS = Mesocortical and mesolimbic systems

TABLE 2.4. Adaptive Activation of the Behavioral and Physical Response to Stress

a.   Central Nervous System Functions
  • Facilitation–arousal, alertness, vigilance, cognition attention, aggression
  • Inhibition–vegetative functions (reproduction, feeding, growth)
  • Activation of counterregulatory feedback loops

b.   Peripheral Functions
  • Increased oxygenation–nutrition of brain, heart, skeletal muscles increased cardiovascular tone, respiration, increased metabolism (catabolism, inhibition of reproduction and growth)
  • Increased detoxification of metabolic products, foreign substances
  • Activation of counterregulatory feedback loops (includes immuno-suppression)

CNS = Central nervous system

The stress system interacts with, influences and is influenced by several systems in the brain subserving cognitive/executive, fear/anger, and reward functions forming with them complex, highly integrated positive and negative feedback system loops (Chrousos 1992, 2009) (Fig. 2.F.4, left). The stress system in a temporally limited fashion activates the central nucleus of the amygdala, which has its own CRH peptidergic system involved in the generation of fear/anger, while reciprocally, the central nucleus of the amygdala stimulates the stress system, forming with it a second important positive reverberating feedback loop. The stress system also acutely and transiently activates the mesolimbic dopaminergic reward system (ventral tegmental area to *nucleus accumbens*) and the mesocortical dopaminergic system of the frontal/prefrontal lobe (F/PFL), while it receives inhibitory input from the latter. Finally, the stress system acutely activates the hippocampus, while it receives from it mostly gamma-aminobutyric acid (GABA)-ergic inhibitory input, some of it as part of negative regulatory feedback from the circulating glucocorticoids of the HPA axis to its hypothalamic center, the paraventricular nucleus, and partly as tonic inhibitory input upon the stress system.

Activation of the stress system stimulates arousal and suppresses sleep (Chrousos 2007). Conversely, loss of sleep is associated with mild inhibition of the stress system, but concurrent elevations of circulating IL-6, a cytokine with fatigogenic and somnogenic properties (Papanicolaou, Wilder et al. 1998).

The growth, reproductive and thyroid axes are inhibited at several levels by stress mediators, while estradiol and triiodothyronine stimulate the stress system (Chrousos 2009). During stress, the gastrointestinal system is inhibited at the level of the stomach via the vagus nerve, while it is stimulated at the level of the large bowel via the LC/NE-activated sacral parasympathetic system. During acute stress, because of catecholamine and cortisol elevations, the heart rate and arterial blood pressure are increased, while gluconeogenesis, glycogenolysis, lipolysis and hepatic glucose secretion are stimulated (Table 2.3).

Stress has complex effects upon the immune system, influencing both the innate and acquired immune response (Chrousos 1995, 2000). Cortisol and catecholamines influence leukocyte and accessory immune cell traffic and/or function and suppress inflammatory cytokines, while both hormones produce a systemic switch from Thelper-1- (cellular immunity) to Thelper-2- (humoral immunity) driven immunity. Inversely, proinflammatory cytokines stimulate the stress system, also at multiple

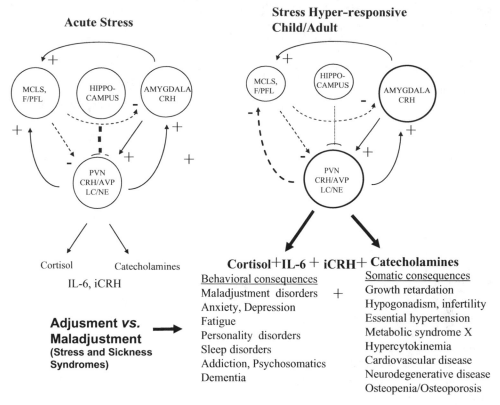

Fig. 2.F.4. Adjustments and maladjustments in the stress response. Acute stress stimulates the stress system (PVN CRH (corticotropin-releasing hormone)/AVP (arginine-vasopressin) and LC/NE), which is in continuous mutual interaction with the amygdala, hippocampus and mesocortical and mesolimbic (MCLS) and frontal/prefrontal cognitive and executive (F/PFL) systems (left). In stress hyper-responsive individuals the interactions of the stress system with the amygdala, hippocampus and MCLS are geared towards a chronic hyper-response in both a quantitative and temporal sense (right). Note that the normally positive input of glucocorticoids to the MCLS becomes negative in chronic stress. Individuals with a genetic predisposition or because they were exposed to early life stress may develop a basally hyperactive and/ or hyper-responsive stress system, as shown on the right with major behavioral and somatic consequences, as summarized in the bottom right. PVN = paraventricular nucleus; LC/ NE = locus coeruleus/norepinephrine. From Chrousos 1998, modified.

levels, in both the CNS and the periphery, causing elevations of glucocorticoids and consequent suppression of the inflammatory reaction.

Through secretion of IL-6, and possibly other immune molecules, the stress system provokes a mild "sickness syndrome" (Chrousos 1995, 2000). This is an innate program of the organism that normally unfolds during a systemic inflammatory reaction, and includes somnolence, fatigue, nausea, and depressive affect, and occurs concurrently with activation of the acute phase reaction by the liver and stimulation of the sensory afferent nervous system manifest as hyperalgesia and fatigue.

The peripheral secretion of "immune" CRH by postganglionic sympathetic nervous system neurons and the norepinephrine-activated release of IL-6 by peripheral immune and other cells, respectively lead to degranulation of mast cells in several tissues (Chrousos 1995, 2000). This represents an important component of the "neurogenic" inflammatory response.

3. Pathological Effects of Stress

In human beings, the crucial beneficial homeostatic or stress mediators that are activated to reestablish homeostasis, including CRH, epinephrine, norepinephrine and cortisol, are also, paradoxically, to a great extent responsible for the damage the organism sustains when in cacostasis (Chrousos 2009; Chrousos and Gold 1992). These mediators may impair the physiology of our cells, disturb their metabolic activity, increase the activity of their inflammatory signaling system and stimulate their oxidative functions, all potentially detrimental changes, ultimately causing atherosclerosis and cardiovascular disease and increasing the chance of the organism to get cancer. In fact, chronic cacostasis has been associated with loss of telomere length, an index of accelerated aging (Al-Attas, Al-Daghri et al. 2010).

Appropriate responsiveness of the human stress system to stressors is a crucial prerequisite for a sense of well-being, adequate performance of tasks, and positive social interactions, and, hence, for the survival of both the self and the species (Chrousos 2009; Chrousos and Gold 1992). By contrast, inappropriate over-responsiveness or under-responsiveness of the stress system may impair growth and development, and may account for the many chronic behavioral, endocrine, metabolic, and autoimmune disorders that plague contemporary humanity. The development and severity of these conditions primarily depend on the genetic vulnerability of the individual, the exposure to adverse environmental (especially social) factors and the timing and duration of the stressful event(s).

Timing of exposure to stressors is particularly effective prenatally and in early life (Chrousos 1998) (Fig. 2.F.5). Indeed, early life experiences may shift an individual's stress system baseline activity and/or responsiveness to the left (lower) or right (higher) levels. Early life programming such as this during the fetal and childhood periods may be associated with epigenetic changes imprinted permanently upon the genome as well as altered neural circuits in the brain.

Through its mediators, stress can produce both medical and psychiatric pathology acutely or chronically upon a vulnerable genetic, constitutional and/or epigenetic background (Chrousos 2009; Chrousos and Gold 1992) (Table 2.5). Acutely, stressors may trigger allergic manifestations, such as asthma, eczema or urticaria, angiokinetic phenomena, such as migraines, hypertensive or hypotensive attacks, different types of pain, such as headaches, abdominal, pelvic and low-back pain, gastrointestinal manifestations from the stomach or gut (pain, indigestion, diarrhea, constipation), as well as panic attacks and psychotic episodes (Table 2.5, left). Chronically, stressors may cause behavioral/neuropsychiatric manifestations such as anxiety, depression, and executive/cognitive dysfunction; cardiovascular manifestations such as hypertension; metabolic disorders such as obesity and the metabolic syndrome including type 2 diabetes; cardiovascular atherosclerotic disease; neurovascular degenerative disease; osteopenia and osteoporosis; and sleep disorders such as insomnia or excessive daytime sleepiness and fatigue (Table 2.5, right).

The pathogenesis of acute stress-induced disorders can be explained on the basis of an increase in the secretion and effect of the major stress mediators upon a vulnerable background (Fig. 2.F.3). Thus, acute allergic attacks may be activated by

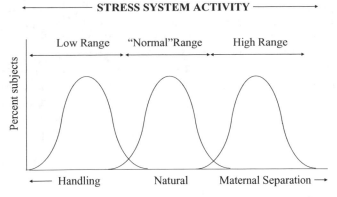

STRESS SYSTEM ACTIVITY

Fig. 2.F.5. Early life experiences and the stress response. Early life experiences may alter an individual's stress system baseline activity and/or responsiveness to the left (lower) or right (higher) levels. In this heuristic schematic diagram, a population of individuals exposed to background "normal" levels of stress as infants is shown in the middle; individuals exposed to handling and comfort as infants have shifted to the left, while individuals who were exposed to the stress of maternal separation as infants are shifted to the right. Early life programming such as this, including the fetal and childhood periods, may be associated with epigenetic changes imprinted upon the genome as well as permanent alterations of neural circuits in the brain.

TABLE 2.5. Acute and Chronic Stress System Pathologies

Acute effects	Chronic effects
• Asthma, eczema, urticaria	• Anxiety, depression, executive and cognitive dysfunction
• Migraines	• Hypertension
• Hyper-/hypotensive attacks	• Obesity/metabolic syndrome
• Myocardial infarction/stroke in compromised hosts	
• Acute pains: headache, abdominal, pelvic, low-back pains	• Cardiovascular atherosclerotic disease
• Gastrointestinal pain	• Neurovascular degenerative disease
• Panic attacks	• Osteopenia and osteoporosis
• Psychotic episodes	• Sleep disorders
	• Immune dysfunction increasing vulnerability to certain infections and allergic/autoimmune disorders

degranulation of mast cell iCRH secreted at the site of the vulnerable organ, e.g., lung for asthma, skin for eczema, etc. Similarly, migraine headaches could be caused by iCRH-induced degranulation of meningeal blood vessel mast cells, causing local vasodilatation and increased blood brain barrier permeability; panic or psychotic attacks could be triggered by central amygdala CRH bursts activating a fear response; hypertensive or hypotensive attacks could be caused by, respectively, stress-induced excessive sympathetic or parasympathetic system outflow. Similarly, acute stress in a compromised host could result in myocardial or cerebral ischemia or hemorrhage and death.

The pathogenesis of chronic stress-related disorders also can be explained on the basis of an increase in the chronic secretion and effects of the major stress and sickness syndrome mediators influencing the activities of multiple homeostatic systems (Chrousos 2009; Chrousos and Gold 1992) (Table 2.5, right; Fig. 2.F.3; Fig. 2.F.4, right). These disorders, thus, represent the chronic maladaptive effects of two physiologic adaptation programs—stress and inflammation—whose mediators are meant to be secreted in a quantity and time-limited fashion, and which have gone askew.

The behavioral consequences of chronic stress result from continuous or intermittent activation of the stress and sickness syndromes and the prolonged secretion of their mediators (Chrousos 2000, 2009; Chrousos and Gold 1992) (Fig. 2.F.4, right). Thus, CRH, norepinephrine, cortisol, and other mediators activate the fear system, producing anxiety and/or anorexia or hyperphagia, the same mediators cause tachyphylaxis of the reward system, producing depression and food, substance or stress craving, while they suppress the sleep system causing insomnia, loss of sleep and daytime somnolence. On the other hand, Il-6 and other mediators, possibly in synergy with ones mentioned above, generate fatigue, nausea, headaches, and other pains. The executive and cognitive systems are also malfunctioning as a result of the prolonged activation of the chronic stress and sickness syndromes, and people may perform and plan suboptimally and make and pursue the wrong decisions. A vicious cycle is generated and sustained, with behavioral maladjustment leading to psychosocial problems in the family, peer group, school, and/or work, sustaining or causing further mediator changes and behavioral maladjustment. In contrast to the adult brain, the young, developing brain is particularly vulnerable because of its increased plasticity and because it lacks prior useful experiences to refer to.

The somatic consequences of continuous or intermittent activation of the stress and sickness syndromes are equally or more devastating (Fig. 2.F.4, right). In developing children growth may be suppressed as a result of a hypofunctioning growth axis; in adults, stress-induced hypogonadism can be expressed as loss of libido and/or hypofertility, and sympathetic system hyperactivity can lead to essential hypertension. Chronic stress mediator hypersecretion, given a vulnerable background and a permissive environment, may lead to visceral fat accumulation as a result of chronic hypercortisolism, reactive insulin hypersecretion, low growth hormone secretion and hypogonadism (Chrousos 2009; Chrousos and Kino 2009) (Fig. 2.F.6). The same hormonal changes lead to sarcopenia, osteopenia and/or osteoporosis. Visceral obesity and sarcopenia are associated with metabolic syndrome manifestations such as dyslipidemia (mainly elevated total cholesterol, triglycerides, and LDL cholesterol, and decreased HDL cholesterol), hypertension, and carbohydrate intolerance or diabetes mellitus type 2. Genetically or constitutionally vulnerable women of reproductive age may develop polycystic ovary syndrome. Stress-related IL-6 hypersecretion plus adipose tissue-generated inflammatory hypercytokinemia, as well as hypercortisolism, contribute to increased production of acute phase reactants and blood hypercoagulation. Together, insulin resistance, hypertension, dislipidemia, hypercytokinemia, and blood hypercoagulation lead to endothelial dysfunction and consequent atherosclerosis with its cardiovascular and neurovascular sequelae.

Chronic stress-induced immune dysfunction, primarily the Thelper1 to Thelper 2 switch, is increasing the vulnerability of individuals to certain infections and allergic and autoimmune disorders (Papanicolaou, Wilder et al. 1998; Chrousos 1995, 2000).

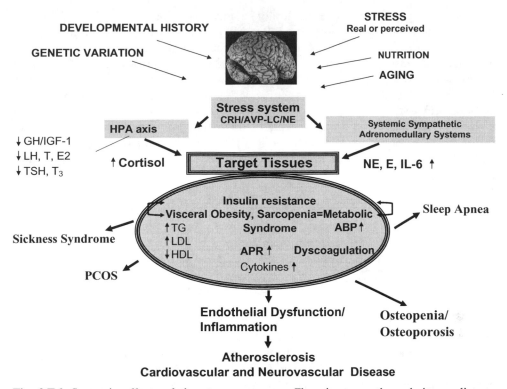

Fig. 2.F.6. Systemic effects of the stress response. Chronic stress, through its mediators, can lead to development of metabolic syndrome and its cardiovascular sequelae. Note also induction of "sickness syndrome" manifestations, polycystic ovary syndrome, sleep apnea. and osteopenia/osteoporosis. CRH = corticotropin-releasing hormone; AVP = arginine-vasopressin; ACTH = corticotropin; F = glucocorticoid cortisol; NE = norepinephrine; E = epinephrine; iCRH = immune CRH; IL-6 = interleukin-6; GH = growth hormone; IGF-1 = insulin-like growth factor 1; LH = luteinizing hormone; T = testosterone; FSH = follicle-stimulating hormone; E2 = estradiol; P4 = progesterone; TSH = thyrotropin; T3 = triodothyronine; TG = triglycerides; LDL = low density lipoprotein cholesterol; HDL = high density lipoprotein cholesterol; APR = acute phase reactants such as C-reactive protein; ABP = arterial blood pressure. From Chrousos and Kino 2009, modified.

Overproduction of CRH and/or stress system abnormalities have been reported in medical, behavioral and neuropsychiatric disorders, such as hypothalamic oligo-amenorrhea, infertility, obligate athleticism, anxiety, depression, post-traumatic stress disorder in children, eating disorders, and chronic active alcoholism (Chrousos 2009; Chrousos and Gold 1992). Overproduction of CRH in the brain and the periphery, and disruption of the HPA axis and the arousal and sympathetic systems, on the other hand, have been reported in obesity/metabolic syndrome and essential hypertension. Furthermore, stress system and autonomic dysregulation is a common feature of gastrointestinal psychosomatic disorders, such as irritable bowel syndrome and peptic ulcer disease.

Neuroendocrine, autonomic, and immune aberrations are also present in chronic inflammatory/autoimmune and allergic diseases, as well as in the chronic fatigue and fibromyalgia syndromes, with considerable evidence linking low CRH activity to the observed abnormalities. Similar low CRH activity has been implicated in atypical, seasonal depression, postpartum blues/depression, and the late luteal dysphoric disorder and climacteric depression (Chrousos 2009; Chrousos and Gold 1992). In all these disorders the problem appears to be cacostasis due to inadequate stress system activity and responsiveness, adversely influencing the functions of several homeostatic systems.

Interestingly, stress system hyperactive/hyperresponsive vs. hypoactive/hyporesponsive states, for instance melancholic vs. atypical depression, respectively, are both associated with development of the metabolic syndrome X and its cardiovascular and other sequelae through two opposed but metabolically converging pathways (Chrousos 2007; Gold, Gabry et al. 2002) (Fig. 2.F.7A). In melancholic depression-like states, we have the standard consequences of chronic stress hormone hypersecretion and, in parallel, chronic insulin hypersecretion, resulting in visceral fat accumulation and sarcopenia, as outlined above. In atypical depression-like states, on the other hand, we have the adverse co-occurrence of chronic sympathetic system hypofunction, resulting in deficient insulin secretion suppression, i.e., *de facto* insulin hypersecretion, a state that also favors visceral fat accumulation, with all that this entails.

Similar to the above, small-for-gestational-age (SGA) fetuses, probably as a result of exposure to stress/undernutrition, and large-for-gestational age ones (LGA) ones exposed to overnutrition, such as those of obese mothers or mothers with gestational diabetes, may also develop metabolic syndrome through similar opposed but metabolically converging pathways (Fig. 2.F.7B).

Modern societies are plagued by the cluster of the so-called multifactorial polygenic noncommunicable disorders that include obesity and the metabolic syndrome with or without diabetes type 2, hypertension, and/or dislipidemia, all leading to cardiovascular and neurovascular disease, as well as autoimmune and allergic disorders, anxiety, insomnia, depression and the so-called "pain and fatigue syndromes" (Chrousos 2004, 2009) (Table 2.6). Every one of these common conditions has been associated with dysfunction of the stress system. This is because the latter has played a crucial role in our survival during our evolution as a species. Indeed, we have survived environmental stressors, which applied selective pressures upon our genome, favoring ancestors who survived because they were efficient at conserving energy, combating dehydration, fighting injurious agents, anticipating adversaries, minimizing exposure to danger and preventing tissue strain and damage. In modern societies, lifestyle has changed dramatically from that of our not-so-distant evolutionary past. The modern environment and the extension of our life expectancy are factors in the expression of these ills.

G. NOTE BY STEFAN BORNSTEIN AND ANDREAS ANDROUTSELLIS-THEOTOKIS ON ENDOGENOUS STEM CELLS AS COMPONENTS OF PLASTICITY AND ADAPTATION

The evolutionary history of humans has been marked by extreme traits, and adaptations to enable them. Humans are bipedal, and to walk on two legs required them to evolve a narrow pelvis. At the same time, their higher cognitive skills

(a)

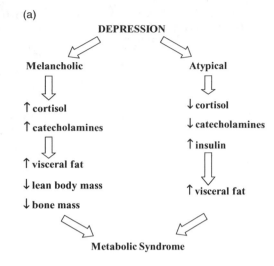

(b)

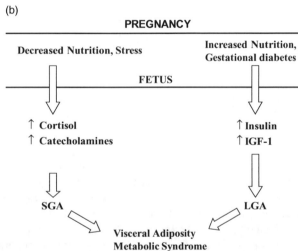

Fig. 2.F.7. From depression to metabolic syndrome. (A) Melancholic and atypical depression representing respectively a stress system-hyperactive/hyperreactive or hypoactive/hyporeactive state are both associated with development of the metabolic syndrome X and its cardiovascular and other sequelae through two opposed but metabolically converging pathways. The central stress system and the peripheral sympathetic system, which normally restrain appetite and insulin secretion, respectively, when hypoactive, allow increased food intake and excessive insulin secretion, both of which promote obesity, including accumulation of visceral fat. From Chrousos 2007. (B) Similar to the above, small-for-gestational-age (SGA) fetuses exposed to stress/undernutrition and large-for-gestational age (LGA) ones exposed to overnutrition, such as those of obese mothers or mothers with gestational diabetes, may also develop metabolic syndrome through similar opposed but metabolically converging pathways. From Chrousos 2007, modified.

TABLE 2.6. Selections of Gene Networks Participating in Functions Important for Human Survival and Species Preservation—All Related to Stress System Activity Changes

Response to Survival Threat	Selective Advantage	Contemporary Disease
Combat starvation	Energy conservation	Obesity/metabolic syndrome
Combat dehydration	Fluid and electrolyte conservation	Hypertension
Combat injurious agents	Potent immune reaction	Autoimmunity/Allergy
Anticipate adversaries	Arousal/fear	Anxiety/insomnia
Minimize exposure to danger	Withdrawal	Depression
Prevent tissue strain/damage	Retain tissue integrity	Pain and fatigue syndromes

From Chrousos 2004, modified.

required a bigger brain, housed in a larger skull (Washburn 1960). As an adaptation to this anatomical problem, much of the brain and head growth takes place after birth.

Postnatal growth patterns in humans provide their own clues to their adaptation abilities. When a child passes from infancy to childhood, growth rates increase, and the timing of this transition partly determines the size of the adult. A delayed childhood will result in a smaller stature, and this has been proposed as a mechanism to adapt to energy crises (Hochberg 2009).

The plasticity exhibited by the developing organism, therefore, directly impacts the traits of the resulting adult. In addition to energy requirements, other environmental and social experiences determine development, including an enriched environment with social stimulation, which help the brain to grow. In this section, we will address how aspects of the plasticity exhibited by the developing organism result in long-term effects, and how they may involve the stem cell compartment that is today accepted to exist in most tissues of the body. We will focus on the brain because it is affected by many factors encountered in childhood (food availability, environment, social interactions, stress), and because many recent animal studies shed light on the mechanism of its plasticity. Additionally, the mammalian brain (developing and adult) is recognized to maintain a pool of stem cells, which can be mobilized by many experiences encountered throughout life, including physical exercise, an enriched environment, and injury). Despite streams of reports that study the involvement of stem cells in the brain's plasticity, we will discuss how the field still lacks an understanding of the stem cell composition of the brain, the role of these cells in development and disease, and the repercussions of manipulating them. These issues, at the moment, raise more questions than they provide answers, but they also bring to light recently discovered populations of cells that may be an integral component of tissue plasticity and the process of adaptation, the resident, or endogenous neural stem cells. Plasticity and adaptation in the central nervous system are not well understood. It is only relatively recently that the roles of stem cells and the generation of new neurons have been implicated in these phenomena. There is still a disconnect between dramatic observations involving changes in the stem cell population in the brain and the determination of consequence. This section has to deal with fairly basic science as general conclusions are, as yet, difficult to form. But the work presented clearly shows that a great deal of new understanding is around the corner, relating to the plastic and adaptive nature of the central nervous system.

1. The Adult Mammalian Brain: Plastic or Rigid?

Relative to other animals, including lower vertebrates, the mammalian brain and spinal cord have very limited regenerative ability. The term "regeneration" is used here to define the reacquisition of lost functions following insults to the tissue. Reacquisition may be due to the replacement of lost cells, their rescue, as well as physiological changes that affect cells and improve their function; alternatively, regeneration may involve changes on nondamaged cells that induce them to compensate for the inflicted damage (e.g., neurons sprouting new axons).

Despite its relatively limited regenerative potential, and in contrast to the decades-long dogma that new neurons are generated only during development in mammals, it is now accepted that the adult mammalian brain is quite capable of generating new neurons, albeit in specific locations. These are the subventricular zone lining the lateral ventricles of the brain, and the subgranular zone of the dentate gyrus in the hippocampus (discussed later). Both these sites of neurogenesis are directly informed by experience: In the case of the subventricular zone, newly generated neurons migrate to the olfactory bulb contributing to the plasticity of olfaction; in the case of the subgranular zone, newly generated neurons are thought to contribute to learning and memory functions (Shors, Miesegaes et al. 2001; Shors, Townsend et al. 2002; Bruel-Jungerman, Laroche et al. 2005; Snyder, Hong et al. 2005; Winocur, Wojtowicz et al. 2006).

Arguably, the mammalian brain is both plastic and rigid. As far as current evidence goes (and we disclaim here that major revisions have happened in the past and may happen again), most of the brain is incapable of replacing lost or damaged neurons, while small areas of the brain do it routinely. One can envision this to be the adaptation to the challenge of learning: A plastic mind can learn, but a rigid mind can store information. Advanced learning and memory requirements necessitate the flexibility to encode new experiences, and this, in turn, requires physical changes to the brain (perhaps including the generation of new neurons). But learned skills require memory to store them, and this should not be easily wiped out. The mammalian brain may have allocated different areas to plasticity and long-term information storage. In the next subsection, we will discuss how our ever-evolving understanding of stem cell biology and neurogenesis in the adult mammalian brain is shaping current thinking on the plasticity of the central nervous system.

2. Hidden Plasticity Potential in the Brain

Historical Overview   In large part, the neural stem cell field evolved in a time when the differences between the young and the adult brain were thought to be fundamental and qualitative. It was perceived that the young brain exhibited neurogenesis as part of its development, whereas the adult brain was completely or largely devoid of this ability. Stem cell scientists used these differences to identify putative stem cells in the developing brain (e.g., fetal rodent brains). Early work involved the generation of monoclonal antibodies against proteins that were highly expressed in the fetal brain and which exhibited little expression in the adult (Hockfield and McKay 1985). Such efforts led to the identification of the first marker of neural stem cells, the intermediate filament protein nestin (Hockfield and McKay 1985; Frederiksen and McKay 1988; Lendahl, Zimmerman et al. 1990).

Using the differences between the developing and adult brains as a source of clues, scientists managed to culture neural stem cells from the fetal rodent brain (Bartlett, Reid et al. 1988; Frederiksen and McKay 1988; Cattaneo and McKay 1990; Deloulme, Gensburger et al. 1991; Lo, Birren et al. 1991; Reynolds and Weiss 1992). Components of the extracellular matrix and cytokines expressed in developing tissues were applied to the culture of neural stem cells, allowing these to proliferate and, when induced to do so, to differentiate into mature progeny (Fig. 2.G.1). The peripheral nervous system proved more challenging from which to establish cultures. However, using cancerous cells from a neuroblastoma cell line, it was eventually shown that dividing, cultured cells can generate active neurons (Schubert, Humphreys et al. 1969; Schubert and Jacob 1970; Lahav, Ziller et al. 1996).

As the field became more successful at identifying and culturing fetal neural stem cells, scientists applied their findings to adult tissues. It was soon realized that stem cells were also present in adult tissues, including the brain, albeit at possibly smaller numbers. Adult neurogenesis, in fact, was first proposed before the identification of

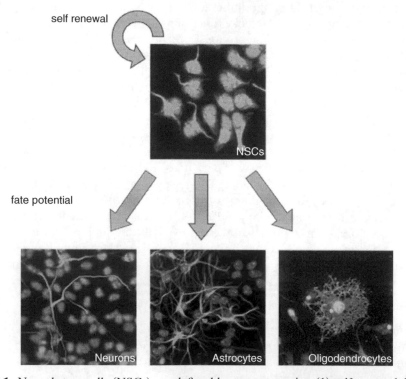

Fig. 2.G.1. Neural stem cells (NSCs) are defined by two properties: (1) self-renewal, i.e., the ability to proliferate and generate more of their own kind of cell, and (2) multipotential, i.e., the ability of their progeny to differentiate into all three major cell types of the nervous system: neurons, astrocytes, and oligodendrocytes (here shown immunostained for the markers beta tubulin, glial fibrillary acidic protein, and CNPase, respectively).

neural stem cells, although it remained a particularly contested notion for many years, partly due to the technical difficulties involved in the experiments required to prove it. The approach involved the use of radioactive thymidine that is incorporated into the DNA of dividing cells. The results showed that most neurons arise from cells that divided only during development. There were exceptions, though: two areas of the adult brain where cells actively dividing in the adult can generate new neurons (Altman and Das 1965; Privat and Drian 1975; Schlessinger, Cowan et al. 1975; Kaplan and Bell 1984; Vaysse and Goldman 1990; Cameron, Woolley et al. 1993; Levison and Goldman 1993; Lois and Alvarez-Buylla 1993; Luskin 1993; Palmer, Ray et al. 1995; Kuhn, Dickinson-Anson et al. 1996; Palmer, Takahashi et al. 1997; Marmur, Mabie et al. 1998) (these are discussed further in the text). Soon after the realization of this newly found plasticity of the brain, it was exploited in the context of degenerative disease, showing that endogenous dividing cells can replace neurons that were lost following damage to the cerebral cortex or striatum striatum (Magavi, Leavitt et al. 2000; Arvidsson, Collin et al. 2002; Parent, Vexler et al. 2002).

One way to understand the impact that insults to the developing brain may have in the long-term, is to understand how insults affect endogenous stem cells in the brain, and what the role is of these cells. The difficulties in answering these questions are obvious when one realizes the lack of consensus in the field as to how many of these cells are present in tissues, what their evolutionary purpose is, and whether they are, in fact, necessary at all, in the context of disease. Before we tackle some of these issues, we need to clarify some confounding views in the stem cell field.

3. Neurogenic Cell vs. Neural Stem Cell

A somatic (tissue-specific) stem cell is defined as a multipotent, self-renewing cell. Self-renewal is the ability to generate more cells of its own kind, and many cells have this ability. In addition, a somatic stem cell must also be able to generate progeny that, when induced to do so (e.g., by pharmacological treatments), will generate all the major cell types of the tissue from which it is derived. For example, a neural stem cell must be able to generate neurons, astrocytes, and oligodendrocytes.

These definitions are largely arbitrary. For example, it is not globally agreed upon for how long in culture a neural stem cell must be able to self-renew to satisfy the self-renewal criterion. Given how finicky these cells are to culture, under the conditions currently used, this is a debated issue, and oftentimes, textbooks suggest that neural stem cells must be able to self-renew "indefinitely." However, in this note, we will avoid using the word "indefinite" as part of a definition.

It is also difficult to find widespread agreement on the second criterion. In the early days of neural stem cell research, if a cell, following several self-renewing divisions, was induced to differentiate (by treating the culture with cytokines, small molecules, serum, etc.) and it was shown that a colony of cells derived by a single putative neural stem cell contained neurons, astrocytes, and oligodendrocytes, that was usually enough to call the founding cell a stem cell. This cell was said to be "tripotent" in that it has the potential to generate three cell types. However, as the criteria bar is being raised, and since we know that there are several subtypes of neurons and glial cells, a new debate has emerged as to whether tripotency is a strong enough standard.

These definitions are important in understanding the mechanisms by which new cells are generated in the body. This, in turn, is important to understand how to manipulate the process that generates new cells in the body. A cell that is capable of self-renewal but that can only generate neurons (not astrocytes or oligodendrocytes), is not a neural stem cell. Instead, it is a neuronal progenitor. Knowing which cell (a stem cell, a neuronal progenitor) generates neurons in a particular context is necessary to choose the right strategy to target it, as these cells may have distinct requirements for their survival, self-renewal, and fate potential.

Neurogenesis is the generation of new neurons following cell division. Neurogenesis does not necessitate the presence of stem cells as a neuronal progenitor is also able to divide and generate neurons. Historically, however, the search for neural stem cells has followed the search for neurogenesis. After all, areas where new neurons are generated are more likely to contain neural stem cells.

4. Does the Role of Neural Stem Cells Change from the Developing Age to the Adult?

It was logical for the search for neural stem cells to focus on the neurogenic areas, since neurogenesis is a property of neural stem cells, although a nonexclusive one. It was also reasonable to assume that the developing central nervous system is richer in neural stem cells than the adult brain, given that stem cells contribute to building tissues. The limited regenerative potential of the adult mammalian brain brought additional pessimism in the quest to find a disease-relevant population of neural stem cells in the adult brain. If the brain does not grow and does not repair, why would there be stem cells in it?

Following neurogenesis, stem cells were found, however, in the adult mammalian brain. These were lining the lateral ventricles of the adult rodent brain, the subventricular zone. In the subventricular zone, new neurons are generated from neural stem cells; as these gradually differentiate into mature neurons, they migrate via the rostral migratory tract to the olfactory bulb, where they replace olfactory neurons. This is an established example of experience-based plasticity involving neural stem cells.

In addition to the subventricular zone, the subgranular zone of the dentate gyrus, an area of the hippocampus, also shows active neurogenesis in adult rodents. These cells appear more restricted in their fate potential properties than actual neural stem cells, but they exhibit a great deal of plasticity as the amount of neurogenesis can be modulated by learning and memory tasks, physical activity, and stress. Despite the clear data showing that many factors affect adult hippocampal neurogenesis, the functional role of this neurogenesis has been difficult to show (Aimone, Deng et al. 2010).

A recent viewpoint in the stem cell field suggests that the transition from the young to the adult brain is accompanied by a change in the role of resident stem cells in the brain and spinal cord. If this is indeed the case, it may provide an additional mechanism to understand why certain treatments like radiation therapy are so much more invasive to the young, where it can lead to cognitive impairments, and may provide ideas to partly consolidate these. Irradiation of rats did show deficits in the neurogenic process in the hippocampus (Monje, Mizumatsu et al. 2002). Specifically, following irradiation, neuronal precursors, which normally physically

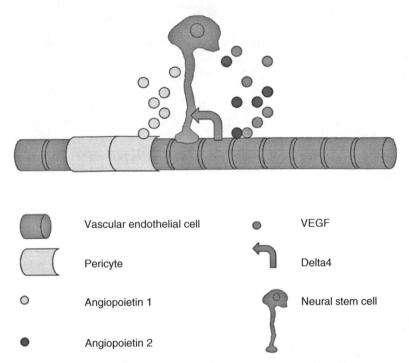

Vascular endothelial cell			VEGF
Pericyte			Delta4
Angiopoietin 1			Neural stem cell
Angiopoietin 2			

Fig. 2.G.2. Neural stem cell response to the environment. The function of the NSCs is subject to the specific environment in which they reside. Their physical association with blood vessels and their expression of receptors for vascular-derived factors makes them particularly sensitive to changes in the environment. Age, disease, and injury-dependent changes in the vascular niche are likely to affect the function of endogenous neural stem cells.

associate with the vasculature of the brain, were found at farther distances from blood vessels, suggesting that they became nonresponsive to the vascular signals that normally regulate their functions (Fig. 2.G.2). Although it is not understood whether this is a causative event for decreased neurogenesis, this study demonstrated a reduction in the number of new neurons in the hippocampus.

Disrupting endogenous neural stem cells in the developing brain will have obvious consequences because these cells are used for the formation of new neurons. It is more difficult to predict consequences in the adult, since our understanding of stem cell function in the mature brain is lacking. Our work has suggested that the adult mammalian brain and spinal cord contain many more neural stem cells than previously thought, and that their distribution is widespread in many areas, including white and gray matter. Previous work has shown that neural stem cells secrete cytokines, such as glial derived neurotrophic factor (GDNF), which are known to promote the rescue of neurons, including dopaminergic neurons, when challenged with toxins (Ourednik, Ourednik et al. 2002). Together, these data suggest a major role of adult neural stem cells in the protection of existing, injured neurons, as opposed to their replacement (e.g., via neurogenesis). Whether these cells are also able to contribute to cell replacement and neurogenesis following injury, remains to be determined.

5. The Disconnect Between Neurogenesis and the Presence of Neural Stem Cells

If adult neurogenesis is confined in the subventricular zone of the lateral ventricles and the subgranular zone of the hippocampus (Altman and Das 1965; Cameron, Woolley et al. 1993; Luskin 1993; Kuhn, Dickinson-Anson et al. 1996; Doetsch, Caille et al. 1999; Bull and Bartlett 2005; Lee, Kessler et al. 2005; Jackson, Garcia-Verdugo et al. 2006), then what is the role of putative neural stem cells outside these two areas? This question is difficult to answer, because of an additional, seeming discrepancy in the role of endogenous neural stem cells: They are easy to activate (e.g., to induce to proliferate and migrate), but it is difficult to prove that this activation is consequential. Endogenous neural stem cells (and precursors; *in vivo* these populations are not always easy to tell apart) can be induced to expand by several types of injury (Liu, Solway et al. 1998; Magavi, Leavitt et al. 2000; Jin, Minami et al. 2001; Arvidsson, Collin et al. 2002; Curtis, Penney et al. 2003; Zhang, Zhang et al. 2008), and through pharmacological stimulation by several soluble factors, including basic fibroblast growth factor (bFGF), epidermal growth factor (EGF), platelet-derived growth factor (PDGF), insulin, perturbation of ephrin signaling, Notch ligands, and the Tie2 receptor ligands, angiopoietin2 (Androutsellis-Theotokis, Rueger et al. 2009; Craig, Tropepe et al. 1996; Kuhn, Dickinson-Anson et al. 1996; Martens, Seaberg et al. 2002; Nakatomi, Kuriu et al. 2002; Holmberg, Armulik et al. 2005; Androutsellis-Theotokis, Leker et al. 2006; Jackson, Garcia-Verdugo et al. 2006; Androutsellis-Theotokis, Rueger et al. 2008; Jiao, Feldheim et al. 2008). In addition, the levels of neurogenesis in the hippocampus can be modulated by physical exercise, experiences within enriched environments, and stress (Schoenfeld and Gould 2011; Kempermann, Kuhn et al. 1997, 1998).

Although most of the work described above focused on the classic neurogenic areas, it is becoming increasingly understood that neural stem cells exist throughout the adult brain and spinal cord, where no evidence for neurogenesis has been conclusively found. Recent studies show that these widespread neural stem cells can be identified by expression of the transcription factor Hes3. In addition, these cells express sonic hedgehog (*in vitro*, sonic hedgehog expression is regulated by Hes3), and the Tie2 receptor ligands (Androutsellis-Theotokis, Rueger, et al. 2010). In fact, Hes3+ cells in the brain express high levels of receptors that can be targeted for their activation, including FGF, Notch, Tie2, and insulin. A case can be made that these cells are plentiful and primed to respond to many types of injury. The debate on the precise roles of these cells still lingers, but, as with many biological systems, it is difficult to imagine that one cell type will have one role. Instead, it is possible, as a means of consolidating these extensive data, that neural stem cells can play complex roles in both cell replacement and neuroprotection.

6. Fetal vs. Adult Neural Stem Cells

The hypothesis that adult neural stem cells play major roles in neuroprotection as opposed to neurogenesis suggests that they may have different properties. These properties may serve as a model to study the transition from the developing to the adult brain. In neural stem cell biology, fetal sources are the most common and best characterized.

Relative to fetal neural stem cells, adult cells are notoriously difficult to culture. However, whether this is an inherent property of the cells or of the environment from which they are isolated (adult vs. fetal brain) is difficult to determine. Apart from this difference, fetal and adult neural stem cells are qualitatively quite similar. Their survival benefits from the same treatments, including bFGF, Notch ligands, angiopoetin2, and inhibitors of p38 MAPK and JAK kinases. In fact, these factors promote the expansion of adult neural stem cells to a greater extent than that of fetal cells. The same conditions induce their differentiation, including serum, and ciliary neurotrophic factor (CNTF). Fetal and adult neural stem cells also express many of the same identifying genes, including nestin, Sox2, and Hes3. These results suggest that fetal and adult neural stem cells are very similar in nature and potential, and that they are controlled by a conserved biology (Johe, Hazel et al. 1996). The possibility arises that the functions and roles of neural stem cells may be determined by the environment in the developing brain. In the next section, we will examine one group of signals that regulate the growth of the organism as well as the expansion of neural stem cells, the insulin-like growth factors (IGFs).

7. Signal Transduction of Stem Cell Regulation

The signals that induce neural stem cells to proliferate and that promote their survival in culture and the living adult brain share many common components with the signals that regulate the growth of a developing child. IGF signaling, a staple of development, is also a major regulator of stem cell survival and proliferation. In addition to the IGFs, other factors including insulin operate via an intracellular signal transduction pathway in stem cells that is common to that downstream of IGFs. Insulin, for example, promotes the survival and expansion of neural stem cells in culture. Omission of insulin from the culture medium can be compensated for by inclusion of other factors such as Notch receptor ligands or IGF1, which intercept the signaling pathways downstream of the insulin receptor (Androutsellis-Theotokis, Rueger et al. 2008). Direct injections of insulin into the lateral ventricles of the adult rat brain induce the expansion of endogenous neural stem cells within days; when these experiments are performed in models of Parkinson's disease, long-term rescue of dopamine neurons is observed (Androutsellis-Theotokis, Rueger et al. 2008). Adult neural stem cells in the rodent brain not only readily respond to insulin activation but they express particularly high levels of the insulin receptor (Androutsellis-Theotokis, Rueger et al. 2009). Although specific studies will have to be implemented to answer this question, these results corroborate the notion that fetal and adult neural stem cells are similar in intrinsic nature, but behave differently at least in part due to the distinct environments they are found in.

Accordingly, while fetal cells are capable of neurogenesis and adult cells in most areas of the brain are not, both behave very similarly when placed in culture, where they exhibit self-renewal and neurogenic fate potential. Possibly, the reasons for the weak neurogenic potential of the adult brain are to be found in the tissue environment and not in the resident stem cell population itself.

To understand how experience, environment, and injury regulate stem cells in tissues, as part of a putative plasticity/adaptation mechanism, it is necessary to understand the signaling pathways that regulate stem cells (Fig. 2.G.3). In this way, we will understand how extracellular signals are interpreted by stem cells and we will be able to predict outcome in the endogenous stem cell population following a variety of stimuli. Recent evidence suggests that neural stem cells, but also human embryonic stem cells and fetal pancreatic precursors, use novel signal transduction

Most mature cells

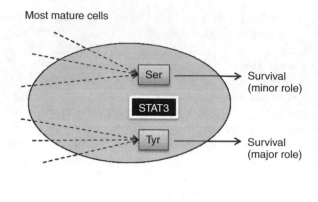

Neural Stem Cells

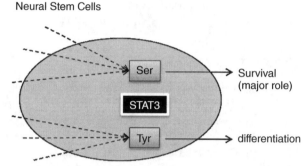

Fig. 2.G.3. Signal transduction in neural stem cells. NSCs have distinct signal transduction requirements. To accommodate their unusual properties (both self-renewal and multipotential), neural stem cells interpret signals in distinct ways from most cells in the body. For example, the signaling molecule STAT3 has two phosphorylation sites: a serine site and a tyrosine site. In most cells in the body, phosphorylation of the tyrosine site promotes their survival, and phosphorylation of the serine site is dispensable. However, tyrosine phosphorylation in neural stem cells is a powerful pro-differentiation signal. Instead, serine phosphorylation is a major determinant of their survival. Multiple extracellular factors lead to serine or tyrosine phosphorylation of STAT3, or both. Understanding the intracellular requirements of stem cells will allow a prediction of the effects of factors during development, injury, disease, and the environment on the stem cell compartment of tissues.

pathways to regulate their numbers, including noncanonical Notch and STAT3 pathways (Androutsellis-Theotokis, Leker et al. 2006). The reason behind this is that stem cells need to accommodate additional cellular properties than most cells within their signaling framework (self-renewal in addition to fate potential). Therefore, these cells may have evolved specific variations of signaling pathways. Elucidation of these signaling pathway "flavors" will allow for better predictions of how environmental signals will affect stem cells.

8. Beyond the Nervous System

Many of the mechanisms that regulate the plasticity of the CNS are shared with other tissues, including cancerous tissues. It is now widely accepted that two critical properties of cancer are carried by a specialized compartment of cells, termed "cancer stem cells." These are recurrence and metastasis. Recurrence necessitates the ability to self-renew (in order to generate the mass of the tumor), as well as fate potential (to generate all the cellular subtypes of the tumor (Reya, Morrison et al. 2001; Shackleton, Quintana et al. 2009). Just like normal neural stem cells, cancer stem cells are capable of both actively proliferating phases and dormancy. This renders them susceptible to different treatments at different times. For example, radiotherapy targets proliferating cells much better than dormant cells. Perhaps someday, our ever-expanding knowledge of what makes a stem cell divide or quiesce will be applied to fight cancer by targeting distinct phases of the cancer stem cell population, the regenerative workhorse of the tumor.

The endocrine system is one of exceptional plasticity and adaptation, as it is on the front line of environmental sensing. In turn, it affects the biology of the entire organism, and the plastic nature of all organs. Endocrine tumors are often benign, despite significant genetic instability (Stratakis 2003), suggesting the possibility of extraordinary control mechanisms to cope with the unusually plastic behavior of this system.

The adrenal gland is a convergence point for various stress sensors, and where coordinated responses are regulated. Cells in this organ are able to proliferate throughout the life span, and the possibility has been suggested that here too stem cells or other immature cells are involved in plasticity (Verhofstad 1993; Chen-Pan, Pan et al. 2002; Powers, Evinger et al. 2007; Ehrhart-Bornstein and Bornstein 2008; Chung, Sicard et al. 2009). Indeed, genes that are also found in immature cells are upregulated in rats subjected to models of stress (Huber, Bruhl et al. 2002; Liu, Serova et al. 2008).

Dysregulation of the neuroendocrine system that involves the adrenal gland leads to both obesity and depression. Drugs for the treatment of these two conditions affect the neuroendocrine system (Licinio, Caglayan et al. 2004; Bornstein, Schuppenies et al. 2006). "Comfort food" may be a self-medication response to deal with chronic stress (Dallman, Pecoraro et al. 2003). Of course, not all obese patients develop depression, and not all depressed patients are obese, suggesting that complex mechanisms regulate the function of the neuroendocrine system.

The realization that the neuroendocrine system contains a plastic population of immature (progenitor) cells offers a new avenue to study the response to stress and the regulation of obesity and depression. It will be critical to study the time constants involved in the progenitor cell of the adrenal gland. This will help us understand how long for a change (even an acute one) in adrenal progenitor cells can affect the biology of the entire organism. It will be particularly important to address the plasticity of these cells, and as a consequence, the sensitivity of the gland during childhood. This will be a starting point to elucidate a mechanism by which changes in the young organism can have consequences for a long time afterwards.

9. Conclusions

There is much to be learned about the role of stem cells in the plasticity and adaptation of tissues to the environment and to injury. Just in the past year, new populations of endogenous stem cells have been identified. The presence of stem cells in adult tissues suggests that some of the mechanisms controlling the developing organism are still present in the adult. The similarities between stem cells from developing and adult tissues, when these are placed in culture (i.e., the same conditions) corroborate this notion. However, the differences in the properties of these cells *in vivo* (young ones readily make neurons, adult ones seem much more restricted), suggest that the local environment of the tissue has a dominant effect on the properties (and role) of these cells. This raises the question of how an insult during development, which may be encoded in the tissue it affects, will impact the resident stem cell population in that tissue, and how this will be translated to the overall state of the tissue. Results from many laboratories show that endogenous neural stem cells are readily regulated by injury, experience, endocrine factors, and pharmacological treatments, showing how much plastic and adaptive potential the CNS carries through its stem cell compartment. That fetal and adult stem cell numbers and function can be regulated by signaling pathways that also regulate the growth of the organism is a formidable clue to understanding the role of these cells during development, and to start manipulating them experimentally and therapeutically.

Acknowledgment for Section G

This work was supported by a grant from the Deutsche Forschungsgemeinschaft KFO 252/1 to SRB and AA-T.

3

FETAL GROWTH

Fetal growth is more rapid in humans than in gorillas or chimpanzees, and both mother and offspring store exceptional amounts of fat (Kuzawa 1998), probably to support an equally exceptional rate of extensive brain growth during the first five years of life.

The first trimester of pregnancy is the time during which organogenesis takes place and tissue patterns and organ systems are established. In the second trimester the fetus undergoes major cellular adaptation and an increase in body size, and in the third trimester organ systems mature ready for extrauterine life. The latter third may be regarded as the first stage of infantile growth (Niklasson and Albertsson-Wikland 2008) (Fig. 3.1), as discussed in the next chapter.

A. ENDOCRINE AND METABOLIC CONTROL OF FETAL GROWTH

Growth hormone from fetal pituitary origin is detected in fetal circulation by week 10 of gestation to be secreted in a pulsatile pattern and under a hypothalamic control. Yet the importance of the fetal growth hormone is poorly established; children with growth hormone deficiency are born only slightly shorter. Moreover, circulating insulin-like growth factor-I in the fetus is independent of fetal growth hormone secretion (de Zegher, Bettendorf et al. 1988). Most investigators in this field feel that the main action of the fetal growth hormone might be its lipolytic and insulin-antagonizing activities. Interestingly, babies who are small for their gestational age have high growth hormone levels that may be the direct effect of reduced feedback drive by low circulating insulin-like growth factor-I levels. Fetal insulin-like growth factor-I concentration is regulated mostly by the energy balance and

Evo-Devo of Child Growth: Treatise on Child Growth and Human Evolution, First Edition.
Ze'ev Hochberg.
© 2012 Wiley-Blackwell. Published 2012 by John Wiley & Sons, Inc.

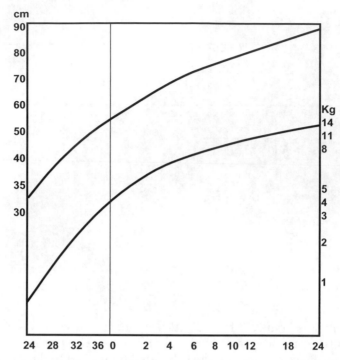

Fig. 3.1. Perinatal growth. The infantile component of growth starts at the 24th week of gestation. The mean length and weight of males show the continuous pattern of growth from gestational age 24 weeks through postnatal 24 months. Data from Niklasson and Albertsson-Wikland (2008).

insulin; low insulin-like growth factor-I in fetuses that are small for their gestational age seems to be caused by undernutrition and low insulin concentration. More importantly, insulin-like growth factor-II and insulin-like growth factor binding protein-3 play a role in fetal growth.

Fetal levels of leptin start increasing around the 34th week of gestation and are correlated with fetal weight (McMillen, Muhlhausler et al. 2004). It has been suggested that leptin is involved in stimulation of fetal bone growth and ossification and lipid catabolism (Ogueh, Sooranna et al. 2000).

B. THE ROLE OF THE PLACENTA

The placenta evolved in eutherian mammals to provide nutrients for the developing fetus, and fetal growth and survival depend on its integrity. Fetal growth and survival depend on placental facilitation of gaseous, nutrient, antibody, and hormonal exchanges, and the disposal of fetal waste products. In order to fulfill its main physiological role as a nutrient sensor and supplier, the placenta follows a carefully orchestrated developmental cascade during gestation (Gallou-Kabani, Vige et al. 2007). Disruption of this cascade can lead to abnormal development of the placental vasculature and/or trophoblast. The time at which some adverse incidents occur during development can have a detrimental consequences on placental function and fetal programming (Myatt 2006). From Insulin-like growth factor-II knockout mice, evidence is emerging that imprinted genes and insulin-like growth factor-II have central roles in controlling both the fetal demand for, and the placental supply of, maternal nutrients (Reik, Constancia et al. 2003). These notions are interesting because deregulation of nutrient supply and demand affects fetal growth, and has long-term consequences for the health of the progeny in the neonatal period and adulthood as a result of fetal programming.

While fetal pituitary growth hormone may have a limited function, placental growth hormone and placental lactogen (hPL), both secreted by the placenta into the maternal circulation, may have an important impact on fetal growth mostly through maternal metabolism and provision of energy from mother to fetus (Fleenor, Oden et al. 2005). Thus, fetal weight at mid-gestation and at term, as well as insulin level, correlate positively with maternal and fetal plasma hPL concentrations (Hochberg, Perlman et al. 1983), and birth weight correlates with maternal placental growth hormone levels. On the other hand, children presenting with mutations of these specific genes do not show significant growth restriction, yet have abnormal biochemical phenotype (Simon, Decoster et al. 1986).

C. DEVELOPMENTAL ORIGINS OF HEALTH AND ADULT DISEASE (DOHaD)

Interest in developmental plasticity in relationship to human health arose from the results of epidemiological and later clinical and experimental studies in which a relationship between early cues (measured often by the surrogate marker of birth weight) and the later risk of metabolic and other diseases was identified (Forsdahl 1977; Barker, Osmond et al. 1989; Barker 1992a, 1992b, 1992c; Barker 2006). This relationship is the basis of the DOHaD hypothesis, and there is growing consensus that this association is broader rather than limited to only those of grossly disturbed early growth. Indeed relationships between the maternal state and later phenotypic changes of pathophysiological relevance can be demonstrated independent of birth weight (Gale, Javaid et al. 2007). The inevitable association of immediate and predictive adaptive responses best models the original birth weight–disease relationships.

In 1986, Barker and Osmond reported a relationship that existed between neonatal mortality in the 1920s and 1930s and the mortality rates from strokes and cardiovascular diseases in the 1960s and 1970s from the same geographic areas (Barker and Osmond 1986). From these epidemiological results, they concluded that the health of the mother and fetus was important in determining the risk of stroke in her offspring, and proposed that cardiovascular diseases might originate during fetal life or early childhood. The results of subsequent epidemiological studies have confirmed this association, and have also shown that small-for-gestational age subjects have a higher risk of hypertension, hypercholesterolemia, impaired glucose tolerance, type 2 diabetes mellitus, and obesity (Barker 1995). Moreover, a poor intrauterine environment that is associated with an unbalanced maternal diet or body composition, placental insufficiency, or endocrine factors induce a phenotype in the offspring that is characterized by an increased risk of developing chronic noncommunicable diseases, such as cardiovascular disease and metabolic syndrome later in life (Godfrey and Barker 2001).

Since the publication of these studies and those of others, the role of fetal or prenatal programming as a determinant of adult diseases has become increasingly clear. Prenatal programming is now known to be an important underlying feature of many systemic adult diseases that includes coronary heart disease, hypertension, insulin resistance syndromes, and osteoporosis, and has direct links to the developmental origins of health and disease hypothesis.

According to this hypothesis, an altered intrauterine environment results in developmental adaptations in the fetus that induce permanent changes in its metabolism and physiology and predispose the individual to cardiovascular disease and metabolic disease in later life. The fetus adapts to the prenatal environment by making permanent adjustments in homeostatic systems that promote its survival in the postnatal environment. This association between poor intrauterine growth and increased risk of disease in later life is the result of a predictive adaptive response where the fetus responds to environmental cues during development with permanent adjustments in its development and homeostatic systems to aid later survival and reproductive fitness. However, an increased risk of disease occurs when the adaptations are inconsistent with the subsequent postnatal environment; they can lead ultimately to an increased risk of disease because its homeostatic capacity is mismatched to that environment (Gluckman and Hanson 2007b).

The association between poor intrauterine growth and increased risk of disease in later life has been replicated in controlled animal-based studies. In these studies, the intrauterine environment is manipulated in order to provide insights into the molecular, cellular, and systemic mechanisms that contribute to the several different manifestations of fetal programming. For example, restricted nutrition during pregnancy in rodents induces dyslipidemia, obesity, hypertension, hyperinsulinemia, and hyperleptinemia in the offspring (Gluckman and Hanson 2007a). From the results of such studies, it is clear that fetal programming is not simply the effect of the passage of heritable genes, but involves the response of the fetus to a variety of intrauterine challenges that trigger lifelong gene expression patterns during critical windows of development. Furthermore, these phenotypes may pass on their effects to future generations in some instances.

The mechanism by which cues about nutrient availability in the postnatal environment are transmitted to the fetus and the process by which different stable phenotypes are induced are unknown, but they are beginning to be understood. The fact that the phenotypic alterations in the offspring persist throughout life suggests the occurrence of stable changes in gene expression.

The DOHaD phenomenon is an example of the broader biological phenomenon of developmental plasticity through which alternative phenotypes (morphs) are generated from a specific genotype in response to persistent environmental cues by adjusting the developmental program (Bateson, Barker et al. 2004; Gluckman and Hanson 2007a). Such phenotypic variation is considered anticipatory of later conditions and represents a predictive adaptive response induced by the organism with the expectation of reproductive fitness benefit.

Gluckman and Hanson postulated that "mismatch" arises when our evolution as a species and our development as an individual do not leave us well-matched to an "evolutionary novel" world (Gluckman and Hanson 2007b). Metabolic disease is an example of this mismatch. The individual variation in the sensitivity of mismatch can be explained in part by genomic variation, and in part by developmental plasticity. We have yet to fully understand the overnutrition pathway, and medicine is reengaging with this development. Nevertheless, a new developmental synthesis is evolving, and this new developmental synthesis must give balanced weight to gene, development, and the intergenerational, past, and current environments in disease causation.

D. IMPRINTED GENES AND INTRAUTERINE GROWTH

The key participants in the growth pathways are regulated epigenetically, as discussed in Chapter 10, Section B, and parent-specific genomic expression—the so-called imprinting—is essential for mammalian ontogenesis. Imprinting is an important genetic mechanism in mammals because imprinted genes can affect intrauterine and postnatal growth, and behavior after birth. Somatic maintenance of imprints throughout development is a highly complex process that involves allelic DNA methylation at imprinting control regions (ICRs), as well as covalent histone modifications and nonhistone proteins (Hochberg, Feil et al. 2011).

Errors in the mechanisms for resetting and maintaining the genomic imprint lead to imprinting defects with or without nucleotide sequence abnormalities (Jiang, Bressler et al. 2004). In humans, dysregulation of imprinting mechanisms has been linked to altered viability, fetal and postnatal growth, neurological development, and behavior. Imprinting occurs at genes for several key hormones, such as the insulin-like growth factor-II and insulin-like growth factor receptor genes that are involved in embryonic and fetal growth. The two human growth disorders, Silver-Russell (intrauterine growth retardation)) and Beckwith-Wiedemann syndromes (intrauterine growth enhancement) are two suitable models to decipher the role of genomic imprinting in growth because specific epigenetic and allele-specific marks are made at imprinted loci, and therefore they can potentially be used as biomarkers for these two diseases (Gaston, Le Bouc et al. 2001). In Beckwith-Wiedemann syndrome, loss of methylation at a centromere region and gain of methylation at a telomere region occurs on the maternal allele. The latter epigenetic defect is associated with a higher risk of cancer in individuals with Beckwith-Wiedemann syndrome (Gaston, Le Bouc et al. 2001).

Aberrant imprinting at chromosome 11p15 disturbs intrauterine and postnatal growth and development, and it is the cause of several disease syndromes that include transient neonatal diabetes mellitus, the Angelman and Prader-Willi syndromes, and uniparental disomies. These growth disorders are caused by abnormal DNA methylation at the 11p15 imprinted region that encompasses many imprinted genes that include insulin-like growth factor-II (Rossignol, Netchine et al. 2008). Epigenetic changes in the 11p15 region occur also in individuals with Silver-Russell syndrome, and this syndrome could be perceived as a molecular mirror of Beckwith-Wiedemann syndrome (Rossignol, Netchine et al. 2008). In these individuals, loss of the telomere methylation occurs on the paternal allele. These 11p15 imprinting anomalies occur probably in the post-fertilization period because of mosaicism in discordant monozygotic twins with either of the two syndromes. However, the epigenetic disruption could occur also during spermatogenesis and oogenesis, and establishment of the parental imprinting mechanism.

E. NOTE BY ALAN TEMPLETON ON THE EVOLUTIONARY CONNECTION BETWEEN SENESCENCE AND CHILDHOOD GROWTH AND DEVELOPMENT

Why do we grow old? Such a question may seem out of place in a book on early development and childhood, yet, there is a strong evolutionary connection between early life and senescence. To understand why, some basic life history theory must first be presented.

1. An Evolutionary Theory of Aging

Life-history theory uses age-specific measures to quantify the ability of an individual to reproduce throughout his or her life. To reproduce, an individual must first be alive, and this is quantified by ℓ_x, the probability that an individual survives to age x given that the individual was alive at age 0. In humans, age 0 is typically regarded as birth, but it can be earlier if fetal mortality is included. However, the standard convention of age 0 being birth is used throughout this note. Given that an individual is alive at age x, then m_x is the expected number of children born to that individual in the age interval described by x. The age interval can be a year or any other time period, depending on the study and the level of temporal resolution required. Because humans are divided into two sexes, and each act of reproduction requires input from both parents, the convention is to assign half a child to one parent and the other half-child to the second parent. Hence, in humans, m_x is the expected number of half-children born to an individual alive at age x in the age interval described by x.

A life-history table gives the age-specific life-history components of ℓ_x and m_x for all the measured ages. Table 3.1 is the life-history table of U.S. females as estimated from the 2000 census. Life-history tables can be defined for many cohorts (a population followed from birth through death), including groups of individuals that share a common genotype or a common phenotype.

The evolutionary impact of natural selection is mediated through variation in reproductive fitness of individuals with different genotypes (Templeton 2006). The life-history table can be used as a fundamental descriptor of the age-specific reproductive fitness effects of genotypes or phenotypes. However, it is often cumbersome to measure reproductive fitness by an entire table of values rather than a single number. Accordingly, there are several measures of reproductive fitness that combine the age-specific fitness components over the entire life span into a single measure of overall reproductive fitness. One such measure is the net reproductive rate, R_0, which is the sum of the product $\ell_x m_x$ over all ages and represents the average number of half-offspring born to members of the cohort over their entire lifetime:

$$R_0 = \sum_{x=0}^{max\,age} \ell_x m_x \qquad (3.1)$$

TABLE 3.1. The Life-History Table for U.S. Females Based on the 2000 Census Data

Age Range (years)	Assigned Age x	ℓ_x	$m_x b_x$	$\ell_x m_x b_x$
<1	0.5	1	0	0
1–4	2.5	0.99376	0	0
5–9	7	0.99261	0	0
10–14	12	0.99189	0	0
15–19	17	0.99107	0.14925	0.1479
20–24	22	0.98909	0.22950	0.2270
25–29	27	0.98671	0.26975	0.2662
30–34	32	0.98392	0.21975	0.2162
35–39	37	0.98021	0.11275	0.1105
40–44	42	0.97460	0.02725	0.0266
45–49	47	0.96623	0.00250	0.0024
50–54	52	0.95398	0	0.0000
55–59	57	0.93561	0	0.0000
60–64	62	0.90716	0	0.0000
65–69	67	0.86344	0	0.0000
70–74	72	0.79983	0	0.0000
75–79	77	0.70983	0	0.0000
80–84	82	0.58563	0	0.0000
85–89	87	0.42145	0	0.0000
90–94	92	0.23936	0	0.0000
95–99	97	0.09669	0	0.0000
≥100	102	0.02479	0	0.0000
Sum:				0.9968

The net reproductive rate reflects the overall probability of living long enough to become sexually mature, find a mate, and successfully reproduce, perhaps multiple times if one lives long enough. As such, it is a measure of lifetime reproductive fitness. For example, U.S. females, based on the 2000 census data, had an average of 0.9968 half-offspring over their entire lifetime (Table 3.1). The net reproductive rate also provides the key to answering the question,: Why do we grow old?

At first glance, it seems that natural selection should favor an *ageless* phenotype in which there is no decline of vigor or reproductive output with age. How could it possibly be an evolutionary advantage for individuals to lose their vigor and reproductive capabilities as they age? If natural selection were simply the survival of the fittest, would not the fittest be an ageless individual? Equation 3.1 is a quantitative measure of lifetime fitness, so we can address these questions by an examination of the mathematical properties of the net reproductive rate.

Let us start with a population of ageless individuals who show no senescence over their entire lifetime. Being ageless is not the same as being immortal. Individuals who do not age still can die through accidents, predation, disease, etc. They are ageless in the sense that the chances of dying in an interval of time does not depend upon their age. Let d be the probability of an individual dying in a unit of time. We regard d as being independent of age and a constant throughout the entire lifetime, reflecting the ageless phenotype of the individual. The individual is also regarded as being ageless with respect to reproduction by letting m, the expected number of half-offspring in a time unit, also be independent of age and a constant throughout the entire lifetime. Given these ageless parameters, the probability of an individual living to age x is:

$$\ell_x = \prod_{i=0}^{x}(1-d) = (1-d)^x \tag{3.2}$$

Then, the net reproductive rate of an ageless individual is:

$$R_0 = \sum_{x=0}^{\infty} \ell_x m = m \sum_{x=0}^{\infty}(1-d)^x = \frac{m}{d} \tag{3.3}$$

using the well-known formula for the sum of a geometric series [$s_n = a + ag + ag^2 + \ldots + ag^{n-1} = a(1 - g^n)/(1 - g) \rightarrow a/(1 - g)$ as $n \rightarrow \infty$, where a and g are constants with $-1 < g < 1$].

Suppose a mutation occurs in this ageless population such that the bearers of this mutation senesce and die at age n-1. The net reproductive rate of the mutant individuals is:

$$R_0' = m \sum_{x=0}^{n-1}(1-d)^x = \frac{m}{d}\left[1-(1-d)^n\right] \tag{3.4}$$

For any $d < 1$ (that is, some death occurs from causes unrelated to age), the term $(1-d)^n$ goes to zero as n becomes sufficiently large, and hence the term in brackets in Eq. 3.4 goes to 1. Thus, if n is large enough (depending on d), then $R_0 \approx R_0'$ and the mutation is selectively neutral as measured by the net reproductive rate.

Natural selection is not the only force leading to evolutionary change. All real populations are finite, and therefore there is sampling error associated with producing the gametes that actually get transmitted to the next generation. This gametic sampling error is known as genetic drift (Templeton 2006) and is a powerful evolutionary force in itself. A diploid population of size N produces $2N$ gametes to create the next generation. If μ is the mutation rate to neutral alleles, then every generation, $2N\mu$ neutral mutations are produced. Any single neutral mutation has a probability of $1/(2N)$ of becoming fixed in a finite population. Fixation means that all individuals in the population will come to bear this mutation. The overall rate of neutral substitutions is therefore (Kimura 1968):

$$2N\mu/(2N) = \mu \tag{3.5}$$

Hence, all populations, regardless of size, fix neutral mutations at a rate given by the neutral mutation rate. Ohta (1976) showed that this rate of neutral evolution also occurs for nearly neutral mutations, given that they are sufficiently close to neutrality. Equation 3.4 ensures that some life-history mutations that cause senescence at large n's will be effectively neutral, and hence all real finite populations will inevitably become fixed for such senescent mutations over long periods of time due to genetic drift. The fixation of these nearly neutral mutations destroys the agelessness of the initial population. A population of ageless individuals is therefore evolutionarily unstable because every finite population will inevitably fix such senescent mutations.

The evolutionary instability of agelessness is enhanced by another class of mutations with life-history effects. Suppose, as before, a mutation occurs that kills its bearers at age n-1. However, we now assume that this same mutation increases earlier reproduction from m to m' such that $m' > m$. For example, suppose this mutation is associated with transferring the energy used in maintaining viability after age n-1 to reproduction at earlier ages. This mutation is therefore associated with a pattern of antagonistic pleiotropy because it is associated with traits that have opposite effects on fitness. The net reproductive rate of the individuals with this antagonistic pleiotropic mutant is:

$$R_0'' = m' \sum_{x=0}^{n-1} (1-d)^x = \frac{m'}{d} \left[1 - (1-d)^n \right]$$ (3.6)

As before, the term in brackets in Eq. 3.6 goes to 1 as n increases to large values, so if the age of onset of the deleterious effects of this mutant is old enough, then its net reproductive rate is approximately m'/d, which is *greater* than the net reproductive rate of the nonmutants, m/d. Bearers of this pleiotropic mutant are actually favored by natural selection as long as the deleterious effects have a late age of onset. In this case, our initial ageless population will evolve senescence due to the positive action of natural selection; that is, it is adaptive to senesce.

Consider now another type of antagonistic pleiotropy. Suppose again that a mutation occurs that kills its bearers at age n-1. However, now assume that this same mutation decreases the death rate in earlier ages from d to d' such that $d' < d$. The net reproductive rate of the individuals with this antagonistic pleiotropic mutant is:

$$R_0''' = m \sum_{x=0}^{n-1} (1-d')^x = \frac{m}{d'} \left[1 - (1-d')^n \right]$$ (3.7)

The term in brackets in Eq. 3.7 goes to one as n increases for any $d' > 0$, so if the age of onset of the deleterious effects of this mutant is old enough, then its net reproductive rate is approximately m/d', which is *greater* than the net reproductive rate of the nonmutants, m/d. Once again, senescence will evolve due to the positive action of natural selection, and it is adaptive to senesce.

The theory summarized above shows that senescence will always evolve if mutations can occur that have a late age of onset for their deleterious pleiotropic effects. If the additional property of antagonistic pleiotropy is added such that there are beneficial pleiotropic effects at younger ages, then natural selection actually favors the evolution of senescence. Anagonistic pleiotropy therefore creates an evolutionary linkage between senescence with survival and reproduction at earlier ages.

2. Thrifty Genotypes and Antagonistic Pleiotropy

Do traits exist in humans that are associated with beneficial effects early in life and deleterious effects later in life? A possible example of this type of trait is type 2 diabetes mellitus. Type 2 diabetes is an adult onset alteration in insulin secretion and insulin resistance (that is, cells do not respond effectively to insulin, a hormone responsible for mediating the uptake by cells of glucose from the blood). Adult onset diabetes is one of the more common diseases affecting humanity, with at least 250 million cases worldwide, and it is increasing at an alarming rate (Alper 2000). The symptoms of this disease show much variation but often lead to many serious complications and even death. As a result of its severity and frequency, adult onset diabetes alone accounts for 15% of the total health-care costs in the United States (Diamond 2003). Type 2 diabetes is usually a disease associated with middle-aged and older individuals. Hence, this is a disease that strongly contributes to senescence through an age-related decline in viability that causes a decline in the human ℓ_x values at a rate faster than could be explained with a constant, ageless death rate.

Why would such a deleterious disease be so common in humans? J. V. Neel (1962) suggested a possible answer to this question: the thrifty genotype hypothesis. This hypothesis is based on antagonistic pleiotropy. Pleiotropy is the phenomenon in which a particular genotype influences more than one trait. Neel speculated that type 2 diabetes did have a genetic component; a speculation that has been strongly borne out by modern genetic studies (Bell, Timpson et al. 2011; Prokopenko, McCarthy et al. 2008; Sladek, Rocheleau et al. 2007). Neel further proposed that these diabetic risk genes were pleiotropic and affected traits other than diabetes. Specifically, Neel speculated that the same genetic states that predispose one to diabetes later in life also result in a quick insulin trigger at younger ages before diabetes could be expressed. Such a quick insulin trigger is advantageous when individuals suffer periodically from famines since it would minimize renal loss of precious glucose and result in more efficient food utilization. This trait would be beneficial to individuals of all ages when subjected to famine conditions, but perhaps most strongly in infants and children. In contrast, the deleterious effects of these genes are only manifest later in life. Moreover, diabetes, like almost all traits, is not determined exclusively by genes or by environmental factors, but rather emerges from the interaction between genes and environment (Templeton 2006). When individuals live in an environment with a low-calorie, low-sugar diet, Neel speculated that they would have little risk of developing diabetes even if they had the predisposing genotypes. These environmental conditions would further reduce the fitness impact of diabetes. Only when there is a high-sugar, high-calorie diet, as found in many modern societies, would there be a substantial risk for the expression of the trait of diabetes, and even then these diabetic genes would at most be mildly deleterious because of the late age of onset of most cases of diabetes (Eqs. 3.4 and 3.7).

The thrifty genotype hypothesis leads to the prediction that the individuals most at risk for type 2 diabetes should be those from populations that had famines or consistently low-calorie diets in their recent evolutionary history but are now living in an environment with a high-sugar, high-calorie diet. As shown by Diamond (2003), this risk pattern has indeed been observed in many human populations (Table 3.2). The Pima Indians in Arizona are one such population from Table 3.2. The Pimas were formerly hunter-gathers and farmers who used irrigation to raise a variety of crops, but principally maize. However, they were living in an arid part of the country, and their maize-based agricultural system was subject to periodic failures during times of drought. This was accentuated in the late 19th century when

TABLE 3.2. Age-Adjusted Incidence of Type 2 Diabetes in Various Human Populations*

Population Grouping	Region	Age-adjusted incidence	History/Diet
Europeans	Britain	2%	Famines mostly eliminated for the past several centuries
	Germany	2%	
	Australia	8%	
	USA	8%	
Native Americans	U.S. Hispanic	17%	Famines in recent history
	U.S. Pima	50%	
Pacific Islanders	Nauru (1952)	0%	History of famines/low-calorie diet
	Nauru (2002)	41%	History of famines/high-calorie diet
New Guineans	Rural	0%	History of famines/low-calorie diet
	Urban	37%	History of famines/high-calorie diet
Aboriginal Australians	Traditional	0%	History of famines/low-calorie diet
	Westernized	23%	History of famines/high-calorie diet

*Modified from Diamond 2003.

European-American immigrants diverted the headwaters of the rivers used by the Pimas for irrigation, resulting in widespread starvation. With the collapse of their agricultural system, the surviving Pimas were dependent on a government-dispensed diet that consisted of high-fat, highly refined foods. Currently, among adult Pima Indians, 37% of the men and 54% of the women suffer from type 2 diabetes, one of the highest incidences known in human populations.

Under the thrifty genotype hypothesis, the extremely high incidence of diabetes in the Pima Indians and other populations (Table 3.2) is due to their recent evolutionary history of high mortality from starvation or famine. However, most human populations have experienced some famine over a timescale of centuries, so the thrifty genotype hypothesis also explains why diabetes is so common in human populations in general.

Modern molecular genetic surveys provide additional evidence for the thrifty genotype hypothesis. Equation 3.7 implies that natural selection will frequently favor the mutations that increase risk to diabetes because young ages receive the beneficial effects while the deleterious effects are not manifested until the older ages, and even then only in a high-calorie-diet environment. When natural selection favors a particular mutation, it not only increases the frequency of that mutation, but also the allelic forms of preexisting genetic variants that just happened to be on the same chromosome as the selected mutation when that mutation first occurred. Hence, as natural selection increases the frequency of the diabetes risk mutation, it also increases the frequencies of the genetic variants associated by chance with the risk mutation. This hitchhiking effect diminishes with increasing chromosomal distance from the selected mutation due to recombination (Templeton 2006). A rapid rise in frequency due to natural selection therefore leaves a distinctive signature in the genomic region surrounding the favored mutation. The genomic regions that harbor risk mutations have now been identified (Bell et al. 2011; Prokopenko, McCarthy et al. 2008; Sladek, Rocheleau et al. 2007). The genomic signatures of recent positive selection have indeed been found at many of the genes that are associated with increased risk for diabetes, particularly in those populations with a history of recent famines (Chang, Cai et al. 2011; Fullerton, Bartoszewicz et al. 2002; Helgason, Palsson et al. 2007; Stead et al. 2003; Vander Molen, Frisse et al. 2005). These studies reveal that natural selection can indeed favor a mutation that predisposes one to diabetes at older ages, exactly as predicted from Eq. 3.7.

When Neel first proposed the thrifty genotype hypothesis in 1962, epigenetics was not a major concern for human geneticists. Today we know that environmental factors often affect the long-term expression of many genes in the human genome. Starvation and severe caloric restriction both *in utero* and during early childhood are environmental factors that appear to induce epigenetic changes that influence the risk for diabetes in later life (Martin-Gronert and Ozanne 2010; Simmons 2007; Stöger 2008). This observation does not undermine the thrifty genotype hypothesis, but rather enhances it. Most traits arise from how genotypes respond to environments, and indeed such a gene–environment interaction is necessary for adaptive evolution via natural selection because without an environmental input into the phenotype of reproductive fitness, there can be no adaptation to an environment (Templeton 2006). Among the traits that can adaptively evolve is the trait of epigenetic plasticity itself; that is, a single genotype can give rise to more than one stable phenotype in response to an environmental signal. Natural selection can favor those genotypes that respond to an environmental signal by producing a high fitness phenotype on the average and eliminate those genotypes that respond to an environmental signal with a low fitness phenotype on the average. In this manner, natural selection can shape the trait of epigenetic plasticity (Templeton 2006).

There is evidence that there is genetic variation for the response to *in utero* starvation that predisposes adults to diabetes. Most European populations have not suffered from famines recently (Table 3.2), but there are exceptions. One such exception occurred in part of Amsterdam during World War II, and a cohort of individuals has been identified that experienced famine *in utero* in 1945 and have increased risk for diabetes as older adults (van Hoek, Langendonk et al. 2009). Several polymorphisms that previous studies had revealed were associated with diabetes risk were surveyed genetically in these famine survivors, and it was discovered that a genetic variant at the *IGF2BP2* locus showed an interaction with exposure to famine *in utero*. Studies such as this show how natural selection for thrifty genotypes can also lead to the evolution of epigenetic triggers for diabetes risk.

3. Thrifty Genotypes and Heart Disease

Mann (1998) proposed a variant of the thrifty genotype hypothesis to explain why humans are so prone to coronary artery disease, another manifestation of senescence in humans. Coronary artery disease (CAD) is initiated by injuries to the endothelial lining of the coronary arteries, followed by the deposition of lipids from low-density lipoprotein (LDL) particles. This results in atherosclerotic plaque. As the plaque grows, it restricts blood flow and changes the mechanical characteristics of the artery wall. These events facilitate plaque rupture, which in turn induces clotting and partial or total blockage of the flow of blood to some heart muscle cells. Depending upon the extent and location of the blockages, symptoms range from mild pain to sudden death. CAD accounts for about one-third of total human mortality in Western, developed societies, making it the most common cause of death. Both genetic and environmental factors contribute to this disease (Sing, Stengard et al. 2003). The lateness in life with which CAD typically occurs and the recentness of the environmental situation in which it is common (Western, developed societies) imply that CAD itself is unlikely to have been the direct target of natural selection during human evolution. Rather, Mann (1998) proposed a model with antagonistic pleiotropy such that the same genes that predispose one to CAD late in life have beneficial effects early in life such that they would be favored, as shown to be possible by Eq. 3. 7.

Past human evolution has been characterized by the rapid and dramatic expansion of our brain and cognitive abilities. The development and maintenance of a large brain creates a high demand for cholesterol and other lipids, both as a source of material for the developing brain and for energy for the brain. This need for fats is greatest during the fetal and infant stages of human development when the brain is growing at a rapid rate (Cunnane 2005). Indeed, humans retain the fetal brain growth rates typical of other apes throughout the first year of life after birth. To ensure sufficient fat to sustain brain growth after birth, much fat is laid down upon the fetus during the last trimester of pregnancy, with the result that humans have the fattest babies at birth of all primates. Indeed, human babies have a greater proportion of body fat at birth than any other mammal except some sea mammals that use fat for insulation (Cunnane 2005). The need for fat in the diet both before and after birth is obviously extreme in humans.

Because the diet of early humans had much less fat in it than the current diets of people living in developed countries, selection would favor those genotypes that were "thrifty" in their absorption and production of cholesterol and other lipids (Mann 1998). Indeed, Corbo and Scacchi (1999) specifically argued that the *ApoE e4* allele is associated with such thrifty lipid genotypes. This same allele has a strong association with CAD (Sing, Boerwinkle et al. 1985). When human life span increased and the diet become much higher in fat, the thrifty *e4* genotypes lead to increased risk for CAD and for other lipid-associated maladies such as Alzheimer's disease (Reiman, Caselli et al. 2001) and Parkinson's disease (Zareparsi, Camicioli et al. 2002). Overall, we see that early brain development growth outweighs, through Eq. 3.7, the late age-of-onset deleterious consequences of human lipid metabolism.

4. Why We Grow Old: The Answer

We grow old because we can die. As long as death is possible, agelessness is evolutionarily unstable. The possibility of death is universal, and therefore so is senescence. Nearly neutral senescent mutations will accumulate in all populations as long as the age of onset is sufficiently large relative to the death rate. Senescence is actually adaptive when antagonistic pleiotropy occurs such that younger ages benefit at the expense of older ages. As we have reduced our death rate from infectious diseases, accidents, and other external causes through our cultures and our ability to manipulate the environment, our deaths have become increasingly due to systemic diseases whose risk increases with increasing age: coronary artery disease, diabetes, cancers, etc.

Modern molecular genetic surveys of the genes that contribute to the age-related risk of these systemic diseases frequently reveal a genomic signature indicative of positive selection; that is, the alleles that predispose us to these age-related diseases were favored by natural selection. As the theory and examples given in this note indicate, the evolutionary advantage to these alleles is their beneficial effects at younger ages, particularly for infants and children. Indeed, the primary trait that makes our species unique, our intelligence, probably could not have evolved without also resulting in senescence (Mann 1998). Thus, senescence is very much tied to early development and childhood through the process of evolution by natural selection. It is adaptive for us to grow old for the sake of our children's growth, development, and survival.

4

INFANCY

In humans and other mammals, infancy is characterized by feeding by maternal lactation, deciduous dentition, rapid and decelerating growth, and an early surge of sex hormones, which peak in humans at age one month and wanes more rapidly in boys than in girls. This so-called mini-puberty seems to be crucial to the organization of male sexual behavior in the brains of many mammals. Whereas breast-feeding continues in natural fertility societies for around 30 months, it varies much in affluent societies. In parallel, the infantile growth stage continues for around the same 30 months, but the next growth stage of childhood sets in affluent societies before the age of 1 year, and in natural fertility societies, such as the Kalahari desert bushmen, the Ju!'hoansi San, at 2 ½ years old (Blurton Jones 1987). This will be discussed in great details in Section E of this chapter.

In natural fertility human societies, where no longitudinal growth data are available, the length of infancy may be defined by weaning from breast-feeding and the mother's new pregnancy (Sellen 2001). Indirectly, it can be evaluated by the interbirth interval as a surrogate for the transition age from infancy to childhood (Gawlik et al. 2011). Even though the interbirth interval is variable according to birth order, as shown for the Ju!'hoansi San bushmen, the first interbirth interval being generally shorter than later interbirth interval, it remains a valid parameter for predicting life history. The interbirth interval lengthens as the number of surviving children increases, until the fourth child; after that it does not vary significantly, although it tends to become shorter. For interbirth intervals subsequent to the first, mortality increased markedly as interbirth intervals decreased (Blurton Jones 1987). Thus, the fitness benefit accruing from more births is balanced against the costs of caring for more children.

Evo-Devo of Child Growth: Treatise on Child Growth and Human Evolution, First Edition.
Ze'ev Hochberg.

A. THE REPRODUCTIVE DILEMMA

Evolutionary fitness requires that within their limited years of fecundity, mothers produce as many viable offspring as possible, and that these would survive until they themselves are able to reproduce. In the case of mammals, evolutionary fitness poses a dilemma: The intense energetic demands of lactation and the typical physical attachment to the infant delay a mother's future reproduction (Kennedy 2005). The mechanism is associated with the milk-generating hormone prolactin, suppressing the secretion of gonadotropins, and the milk ejector hormone oxytocin, both now are known to be involved in love and bonding.

In humans, the dilemma is aggravated by even greater energetic demand by rapid infantile brain growth on the one hand, and, on the other, a greater attachment of mother to baby to ascertain a high-quality infancy. Yet weaning a nursing infant too soon places it at a variety of risks, including increased mortality from environmental hazards and morbidity from infectious and parasitic diseases that potentially restrict growth and development.

Thus, the particular timing of weaning provides a trade-off that determines population growth and lifetime female fertility on the maternal side, and on the infant side it determines growth rates and final adult size, as well as the onset and offset of subsequent life-history stages.

The heavy energetic demands of gestation, lactation, and rearing of offspring focus on female reproductive efforts. However, in several mammals, including humans, paternal care and investment evolved. This apparent altruism on the father's side is well invested in promoting his own fitness. The relative cost of reproduction for males and females determines cooperative and competitive strategies. In particular, when male reproductive costs are less than female reproductive costs, the male's cooperation with females improves his evolutionary fitness even when females do not reciprocate (Key and Aiello 2000).

In the next life-history stage of childhood, other family members will assist in child care, including grandparents, other adult females, some adult males, and even juveniles. But during infancy the child is in the sole care of his mother and, to some small extent, the father.

B. THE OBSTETRICAL DILEMMA

Bipedalism of the great apes some 5 million years ago (Fig. 4.B.1), along with growing brains, were considered to be important steps in human evolution. Yet the 3-million-year gap between these events makes an association between them hardly imaginable. In order to walk on two legs, the foramen magnum and the brain stem moved from back to front to allow an upright posture with the face forward, without placing undue strain on the neck or back. The spine and legs realigned, and the pelvis narrowed markedly (Fig. 4.B.2). The pelvic ilium bone shifted forward and broadened, while the pelvic ischial bone was reduced in size, narrowing the pelvic birth canal.

Three million years after bipedalism ensued, a more developed and neurologically more advanced brain required a larger skull, with a consequent discrepancy between the growing newborn skull and the narrowing maternal birth canal. This was described

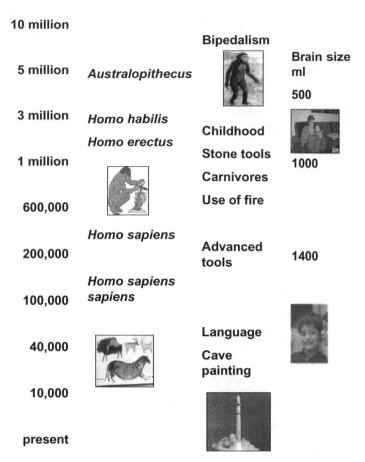

Fig. 4.B.1. Milestones in 10 million years of hominids' evolution. Humankind may be defined by bipedalism, with onset 5 million years ago, growing brain and head as of 2 million years ago, the childhood life history stage, introduced 2 million years ago, and language, appearing 50,000 years ago.

as "the obstetrical dilemma" by Washburn in 1960 (Fig. 4.B.2) (Washburn 1960). He proposed that, as an adaptation to this unfavorable ratio between a smaller maternal birth canal and a larger fetal skull, head and brain growth along with mental and motor development have been deferred into the postnatal period (Leutenegger 1973). Brain growth tapers off after birth in other primates but continues in a rapid, fetal-like trajectory in humans for the first year of postnatal life and at a slower pace thereafter until the childhood–juvenility transition (Fig. 4.B.3). The trade-off for this heterochronicle "dilemma" has been a small-headed, and therefore immature, helpless, and vulnerable newborn. Vulnerability of a helpless being has its own obvious toll on selection, and will be discussed later on. To reach a maturity stage equivalent to those of most other mammals, gestation should have been prolonged to 21–24 months. The 12- to 15-month difference between this figure and the human 9-month gestation becomes an interesting timeframe that will be discussed later—this is the age at which infants in Western societies accelerate their growth after the typical infantile growth deceleration into the childhood phase.

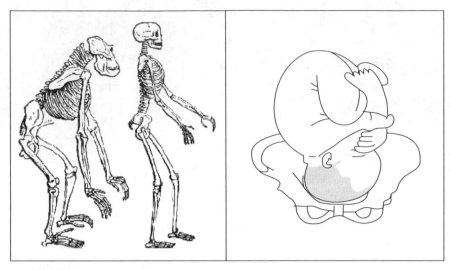

Fig. 4.B.2. The "obstetrical dilemma" of Washburn (1960). Bipedalism of the hominids required realignment of the spine and legs and narrowing of the pelvis, whereas a more developed and neurologically more advanced brain required a large skull. As an adaptation to this unfavorable ratio between a smaller maternal birth canal and a larger fetal skull, head, and brain, growth has been deferred into the postnatal period.

The infancy of humans and other mammalian species are comparable in many respects, such as feeding by maternal lactation and the appearance of deciduous teeth. However, human infancy begins with a much less mature fetus, as growth and brain maturation are deferred into infancy and beyond. Moreover, in most mammals, and all other primates, infancy and lactation end with the eruption of the first permanent molars (Smith and Tompkins 1995). In modern humans, by contrast, there is an interval of about 3–5 years between weaning and the eruption of the first permanent molars at age 6–7 (Fig. 4.B.4).

Because a juvenile mammal's survival depends on having a working dentition, tooth eruption is another important life-history variable. Interestingly, the eruption of primary incisor teeth occurs in modern humans at the same 6- to 12-month interval as that of the infancy–childhood growth transition. In general, tooth eruption has a poor correlation with child growth (Gron 1962), and we do not know whether early teethers are the same children who have an early infancy–childhood transition. We do know that delayed eruption of teeth may be the result of a nutritional problem, and primary teeth appear in children in Lahore, Pakistan, later than in Swedish children (Saleemi, Hagg et al. 1996).

The brain systems that motivate humans to form emotional bonds with others probably first evolved to mobilize the high-quality maternal care necessary for reproductive success in placental mammals (Pedersen 2004). In these species, the helplessness of infants at birth and their dependence upon breast-feeding and parental body heat to stay warm required the evolution of a new motivational system in the brain to stimulate avid and sustained mothering–nurturing behavior. These had to evolve in the infantile brain in addition to maternal behavior in the mother and monogamous bonds between breeding pairs. Maternal and pair bonding

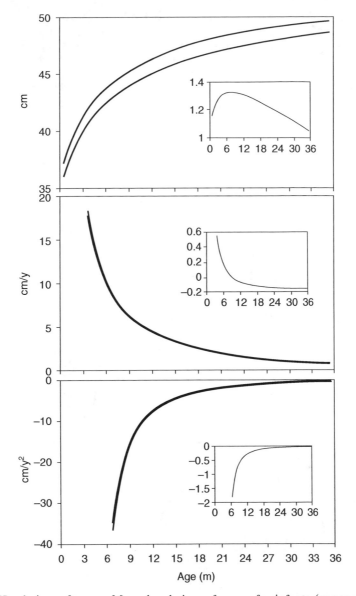

Fig. 4.B.3. Head circumference. Mean head circumference for infants (upper panel), head circumference velocity (cm/y, center) and acceleration (cm/y2, lower panel) for boys (black lines) and girls (gray lines). The first and second derivatives were calculated from Gerver and de Bruin (1996). The insets show the difference between the sexes: males – females.

are accompanied by increased aggressiveness toward perceived threats to the infant as well as diminished fear and anxiety in stressful situations (Pedersen 2004).

While the duration of infancy is shorter in humans than in any other mammal, its quality is highest; in traditional societies, mothers rarely put their babies down, providing breast milk and body warmth, and then do so for no more than a few seconds, usually remaining within a meter of the baby. The quantity and quality of

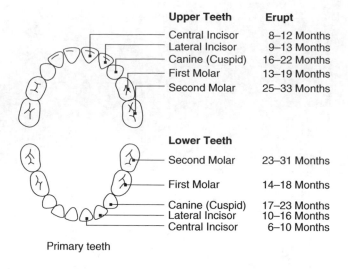

Upper Teeth	Erupt
Central Incisor	8–12 Months
Lateral Incisor	9–13 Months
Canine (Cuspid)	16–22 Months
First Molar	13–19 Months
Second Molar	25–33 Months

Lower Teeth	
Second Molar	23–31 Months
First Molar	14–18 Months
Canine (Cuspid)	17–23 Months
Lateral Incisor	10–16 Months
Central Incisor	6–10 Months

Primary teeth

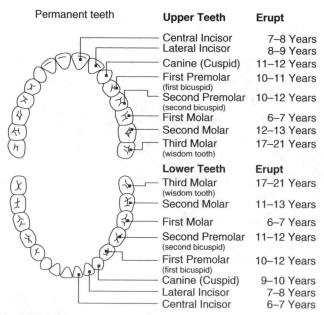

Permanent teeth

Upper Teeth	Erupt
Central Incisor	7–8 Years
Lateral Incisor	8–9 Years
Canine (Cuspid)	11–12 Years
First Premolar (first bicuspid)	10–11 Years
Second Premolar (second bicuspid)	10–12 Years
First Molar	6–7 Years
Second Molar	12–13 Years
Third Molar (wisdom tooth)	17–21 Years

Lower Teeth	Erupt
Third Molar (wisdom tooth)	17–21 Years
Second Molar	11–13 Years
First Molar	6–7 Years
Second Premolar (second bicuspid)	11–12 Years
First Premolar (first bicuspid)	10–12 Years
Canine (Cuspid)	9–10 Years
Lateral Incisor	7–8 Years
Central Incisor	6–7 Years

Fig. 4.B.4. Sequence of primary and permanent teeth maturation. Eruption of primary first molars occurs at the transition from infancy to childhood, and that of the permanent first molar at the transition to juvenility. The second permanent molars erupt at transition to adolescence and the third at the biological age of the first reproduction.

maternal care received during infancy determines adult social competence, ability to cope with stress, aggressiveness, and even preference for addictive substances. Indeed, the development of neurochemical systems within the brain that regulate mothering, aggression, and other types of social behavior, such as the oxytocin and vasopressin systems, is strongly affected by parental nurturing received during infancy. It has been suggested that the neural circuitry and neurochemistry implicated in studies of lower mammals also facilitate primate/human interpersonal bonding, and that neural bonding systems may also be important for the development in individuals' loyalty to the social group and its culture (Pedersen 2004).

Interestingly, the great Swiss developmental psychologist Jean Piaget (1896–1980), who identified the stages of child cognitive development, defined infancy as the "sensorimotor period," from birth to about age 2, when thoughts derive from sensations and movements (Singer and Revenson 1998) (Table 4.1). During this

TABLE 4.1. Piaget's Stages of Cognitive Development

Approximate Age	Developmental Stage & Characteristic Behavior
0–24 months	**Infancy: Sensory Motor Period** At its end, evidence of an internal representational system. Symbolizing the problem-solving sequence before actually responding. Deferred imitation.
2–7 years	**Childhood: Preoperational Period**
2–4 years	**Preoperational Stage:** Increased use of verbal representation but speech is egocentric. The beginnings of symbolic rather than simple motor play. Transductive reasoning. Can think about something without the object being present by use of language.
4–7 years	**Intuitive Stage:** Speech becomes more social, less egocentric. The child has an intuitive grasp of logical concepts in some areas. However, there is still a tendency to focus attention on one aspect of an object while ignoring others. Concepts formed are crude and irreversible. Easy to believe in magical increase, decrease, disappearance. Reality not firm. Perceptions dominate judgment. In moral-ethical realm, the child is not able to show principles underlying best behavior. Rules of a game not developed, only uses simple do's and don'ts imposed by authority.
7–11 years	**Juvenility (Middle Childhood): Concrete Operational Period** Evidence for organized, logical thought. There is the ability to perform multiple classification tasks, order objects in a logical sequence, and comprehend the principle of conservation. Thinking becomes less transductive and less egocentric. The child is capable of concrete problem solving. Class logic-finding bases to sort unlike objects into logical groups where previously it was on superficial perceived attribute such as color. Categorical labels such as "number" or "animal" now available.
11–15 years	**Adolescence: Formal Operational Period** Thought becomes more abstract, incorporating the principles of formal logic. The ability to generate abstract propositions, multiple hypotheses, and their possible outcomes is evident. Thinking becomes less tied to concrete reality. Formal logical systems can be acquired. Can handle proportions, algebraic manipulation, other purely abstract processes. Prepositional logic, "as-if" and "if-then" steps. Can use aids such as axioms to transcend human limits on comprehension.

stage, the child learns about himself and his environment through motor and reflex actions. The child learns that he is separate from his environment and those aspects of this environment—his parents or favorite toy—continue to exist even though they may be outside the reach of his senses. Intelligence is demonstrated in the infant through motor activity without the use of symbols that develop at the end of this stage. Knowledge of the world is limited, but developing, and is based on physical interactions and experiences.

C. GROWTH OF THE INFANT

Infantile growth has been assumed to begin at mid-gestation and to taper off at approximately 2–3 years of age, representing the postnatal extension of fetal growth (Niklasson and Albertsson-Wikland 2008) (Fig. 3.1), and is regarded as being nutrition dependent and closely linked to the action of insulin-like growth factors (Wang and Chard 1992; Leger, Oury et al. 1996). In contrast with nonhuman primates, humans deposit significant quantities of body fat *in utero* and are consequently one of the fattest species on record at birth. Greater adiposity of human neonates is at least partially understandable as complementary to the enlarged human brain, which demands a larger energy reserve to ensure that its obligatory needs are met when the flow of resources from mother or other caretakers is disrupted (Kuzawa 1998). Energy storage in adipose tissue is an important life-history strategy of humans and a means to minimize mortality risk during the nutritionally unstable and complete dependence period of infancy (Kuzawa 1998).

The analysis of infantile linear growth presented in Fig. 4.C.1 is derived from the U.S. National Center for Health Statistics (NCHS) 2000 Centers for Disease Control (CDC) growth charts (CDC 2000). Derivation of the data generates the growth velocity (first derivative) and growth acceleration charts (second derivative) that allow for a meaningful interpretation. Initial growth during the first month of life is at an amazing mean 47 cm/year for males and 43 cm/year for females. Over the initial 9 months of life, it decelerates rapidly (mean -138 cm/year^2 in males and -115 cm/year^2 in females in the second month) to 16 cm/year in both sexes. It will be shown later that the age of 9 months is quite unique as it represents the transition from the infantile to childhood growth stage. Growth steadily and slowly decelerates further after age 9 months for the next 15 months of life, to a mean velocity of 10 cm/year and deceleration of -2 cm/year^2 at 2 years of age for both sexes.

It is remarkable that at as early as the first month of life, sexual dimorphism is evident in infantile growth. The insets of the three charts in Fig. 4.C.1 give the differences between male and female, showing the impact of male against female infantile mini-puberty (see the next chapter). After completion of the male mini-puberty, and while the female's mini-puberty goes on, the male–female differences disappear, suggesting that the female event has no effect on the growth sexual dimorphism, which depends solely on the male mini-puberty.

Comparative auxology shows that the infantile growth pattern is completely different in apes. Gorillas, but also all other mammals, accelerate their infantile growth, as compared to humans' decelerating growth (Fig. 1.D.1). The decelerating stage has happened in the gorilla prenatally, and acceleration will happen in humans in the

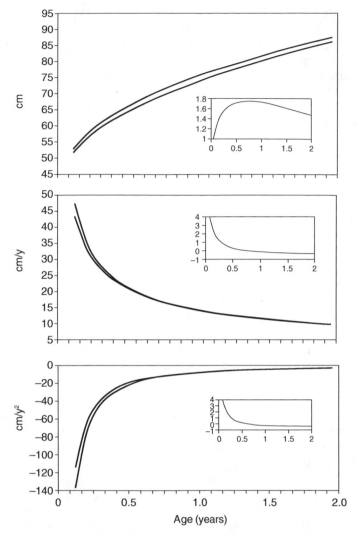

Fig. 4.C.1. Infantile growth. Mean infant length (upper panel), growth velocity (center) and growth acceleration (cm/y², lower panel) for boys (black lines) and girls (gray lines). The first and second derivatives were calculated from the CDC (2000). The insets show the difference between the sexes: males – females; sexual dimorphism is established during infancy.

transition to childhood, indicative of the evolution of the growth pattern. Postnatally, gorillas grow faster than either pygmy chimpanzees or common chimpanzees, continuing a prenatal trajectory (Leigh and Shea 1996). Male growth in the pygmy chimpanzee proceeds at a slower rate than that of the common chimpanzee between the ages of 1 and 4. Adult female common chimpanzees are larger than adult pygmy chimpanzees mainly because they have higher growth rates very early on, and grow for a much longer period. In the female pygmy chimpanzees, more rapid growth later on does not compensate for slower early growth. I will later show that similar to the female pygmy chimpanzee, the transition from infancy to childhood determines adult height in humans.

D. ENDOCRINE ASPECTS OF INFANTILE GROWTH

The unique infantile growth pattern, as described above, is controlled by two endocrine systems: the growth hormone–insulin-like growth factor axis, which gradually gains in importance throughout the youth life-history stages, and the hypothalamic–pituitary–gonadal axis that takes the child through mini-puberty. Growth hormone circulating levels are high in the fetus and newborn compared with those in later childhood and in adults, but its bioactivity is low and insulin-like growth factor axis-I levels are low, resembling a state of growth hormone insensitivity (de Zegher, Kimpen et al. 1990).

During infancy, growth hormone is constantly secreted, lacking the typical pulsatile pattern of later stages. This high basal secretion pattern can be part of the phenomenon of growth hormone insensitivity but it can be due to the slow growth hormone pacemaker maturation. Later in childhood, growth hormone will be released in association with the sleep pattern, as it is produced by the activity of excitatory and inhibitory neurons in several brainstem and forebrain centers. The regions that mature earliest are usually the medulla and pons, followed by the midbrain, thalamus, and hypothalamus, and then by the cerebral cortex and striatum. Yet, during infancy, there is a continuous decrease in total sleep time and REM sleep along with a concomitant increase in sleep efficiency and non-REM sleep (Louis, Cannard et al. 1997). It is during non-REM sleep that the activity of the thalamic reticular nucleus and cortex pathways induce growth hormone release. In non-REM sleep, the inhibitory influence that prevents the spread of arousal activity along the pathways from the brainstem to the cortex is more prominent than in REM sleep. As the infant matures, the frequency of total arousals, cortical arousals, and subcortical activations decrease, affecting GH-inhibitory REM sleep and GH-stimulatory non-REM sleep. This maturation process was found to be similar in breast-fed and bottle-fed infants across the different ages and sleep changes (Montemitro, Franco et al. 2008).

The second characteristic endocrine pattern of infancy is the so-called mini-puberty. The term "mini-puberty" denotes the surge in gonadotropin levels that occurs after birth and peaks at 2 to 3 months (Fig. 4.D.1) (Burger, Yamada et al. 1991). In the male, serum luteinizing hormone (LH) increases beyond female concentrations and in the female, serum follicle-stimulating hormone (FSH) increases more than male concentrations. The hypothalamic–pituitary–gonadal axis is operational during this period, and agonadism at this age, be it part of Turner's syndrome or other conditions, is associated with postmenopausal concentrations of gonadotropins. After peak levels of gonadotropins and sex steroids at 2 to 3 months, they gradually decline over several months to reach a nadir at 6 to 9 months in boys and 2 to 3 years in girls (Winter and Faiman 1973; Burger, Yamada et al. 1991; Grumbach 2005). As the infantile hypothalamus matures, it expresses more sex steroid receptors, increasing the hypothalamic sensitivity to negative feedback by minute levels of sex steroids, while receiving central nervous system inhibitory signals. Thus, the male's mini-puberty is robust, short, and strongly affects infantile growth, whereas the female's mini-puberty is milder, longer, and hardly affects growth.

The concurrence of male, but not the female, hypothalamic–pituitary–gonadal quiescence (Crofton, Evans et al. 2002) with the onset of growth transition from infancy into childhood provides the endocrine definition of this switch. The female

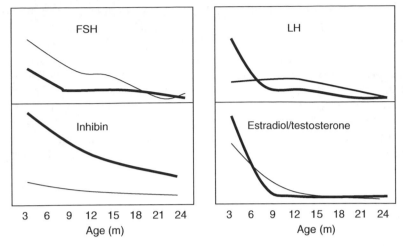

Fig. 4.D.1. Mini-puberty. Age-dependent relative levels of LH, FSH, inhibin, estradiol and testosterone in boys (black lines) and girls (gray lines). Notice the decline in the male hypothalamic–pituitary–gonadal activity toward the age of 9 months, when infantile growth accelerates to childhood growth, and the continuing activity in the female to age 21–24 months. Data from Burger, Yamada et al. (1991).

does not undergo a complete hypothalamic–pituitary–gonadal quiescence; at the end of the life-history infancy stage, serum FSH and estradiol decline, but remain higher than those of the male (Klein, Baron et al. 1994). Biological activities of the hypothalamic–pituitary–gonadal axis are quite apparent during infancy: facial comedones and acne-like lesions may be noticed, infantile mammoplasia is quite common, but more importantly, testicular volume increases due to an increase in seminiferous tubule size, and Sertoli and germ cells proliferate for about 100 days after birth, in parallel with high FSH levels (Muller and Skakkebaek 1983).

Figure 4.D.1 shows the possible impact of this endocrine activity on infantile growth. The sexual dimorphism mentioned above is evident in the length curve, with increasing male-female difference during the initial six to nine months, while testosterone is still high in the boy, and its reversal thereafter, as androgens recede while the female's hypothalamic-pituitary-gonadal axis continues to operate.

E. INFANCY–CHILDHOOD TRANSITION: DETERMINATION OF ADULT STATURE

In the following paragraphs I will present the data and theory that there is an evolutionary adaptive strategy of plasticity in the timing of the transition from infancy into childhood in order to match environmental cues and energy supply (Hochberg and Albertsson-Wikland 2008). We proposed that humans evolved to withstand energy crises by decreasing their body size, and that evolutionary short-term adaptations to energy crises probably utilize epigenetic mechanisms that modify the transition into childhood, culminating in short stature.

A great deal of attention has been paid to intrauterine nutritional "fetal programming" leading to a "thrifty phenotype" and later health consequences, as presented in Chapter 3, Section C. In terms of fetal programming of final height, the vast majority of infants with intrauterine growth retardation catch up immediately after birth and end up normal in stature. Hence, in evolutionary terms, fetal growth, which depends on uterine and placental function, is too labile a stage to play the role of adaptive responder for determination of final height. On the other hand, a delay in the infancy–childhood growth transition is not associated with catch-up growth, indicating that events during the first year of life impact on ultimate height. In analyzing close to 400 children with idiopathic short stature or those born small-for-gestational age and did not catch up early in infancy, we have not encountered a single case that showed a catch-up growth (Hochberg and Albertsson-Wikland 2008).

Pediatricians speak of an infant's failure to thrive (FTT) when his weight or height gains are insufficient. This is observed mostly during the childhood onset period of 6–12 months and beyond. It has been shown that whether the cause was organic or nonorganic, such children often remain short (Drewett, Corbett et al. 1999; Rudolf and Logan 2005), and the mechanism of that delay is delayed infancy–childhood transition (DICT).

Based on the predictive adaptive response theory, the expected response to a secure environment includes the investment in large body size, whereas the expected response to a threatening environment will include a reduction in body size. At the same time, predictive responses to resist threatening and difficult environments may include an altered hypothalamic–pituitary–adrenal axis, altered behavior and anxiety, increased appetite and tendency to store fat, altered food preference, reduced motor behavior and skeletal muscle mass and strength, altered vasculature function, altered insulin release and action, and leptin resistance (Gluckman and Hanson 2007a).

Based on an analysis of growth parameters, the "infancy–childhood–puberty" (ICP) growth model of Johan Karlberg divided human growth into three successive and partly superimposed stages that reflect the endocrine control mechanisms of the growth process (Karlberg 1987; Karlberg, Jalil et al. 1994) (Fig. 4.E.1). While this model ignores juvenility, it is still valid in the argument for the infancy–childhood growth transition. The infancy stage of the ICP model has been assumed to begin at mid-gestation and to tail off at approximately 2–3 years of age, representing the postnatal extension of fetal growth, and is regarded as being nutrition dependent and closely linked to the action of insulin-like growth factors. The childhood growth stage starts in affluent Western countries between 6 and 12 months of age (Karlberg, Engstrom et al. 1987; Liu, Chism et al. 1999; Hochberg and Albertsson-Wikland 2008), and continues through puberty until growth ceases at attainment of adulthood. Thus, the ICP model proposes a period of transition whereby the initiation of the childhood growth stage overlaps with the infancy growth stage and the infantile life-history stage, as defined by weaning in traditional societies at 2–3 years of age. After gradual deceleration of the postnatal infantile growth stage, the growth rate abruptly increases between 6 and 12 months of age (Fig. 4.E.2) (Karlberg 1987; Karlberg, Engstrom et al. 1987; Hochberg and Albertsson-Wikland 2008).

In affluent Sweden, the childhood growth stage begins between 6 and 12 months of age (mean ± 2 SD), with a mean age of 9 months (Karlberg, Engstrom et al. 1987;

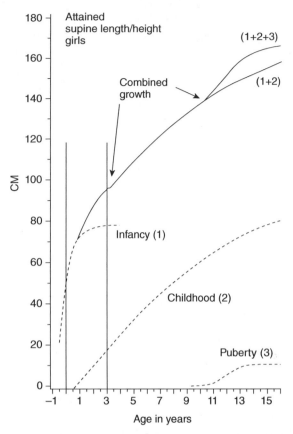

Fig. 4.E.1. The "infancy–childhood–puberty" (ICP) growth model. The model divides human growth into three successive and partly superimposed stages. The infancy stage of the ICP model (here for girls) begins at mid-gestation and tapers off at 2–3 years of age. Childhood sets on at a mean age of 9 months, and puberty at a mean age of 10.5 years. Data from Karlberg (1989) and Karlberg, Jalil et al. (1994).

Liu, Chism et al. 1999). The mean age is 10 months in Israel (Zuckerman-Levin 2007), and 11 months in Shanghai (Liu, Albertsson-Wikland et al. 2000). Whereas about 3% of Swedish infants have an infancy–childhood transition (ICT) beyond the age of 12 months, 7% of Israeli infants show such a delay, and about 11% of the infants in Shanghai fall into that category.

Children with a DICT have both normal infantile and normal childhood growth rates, indicating normally functioning growth control mechanisms during both of these life-history stages (Fig. 4.E.2) (Hochberg and Albertsson-Wikland 2008). Their only abnormal experience is the late transition age. As a result, children with a DICT are longer at birth and during infancy than short children with a normal transition age, while their heights during later life history stages are comparable.

DICT has been reported in as many as 50% of children with "idiopathic" (reason unknown) short stature (Kristrom, Hochberg et al. 2007; Hochberg and Albertsson-Wikland 2008). This effect is time dependent, with a longer delay resulting in shorter

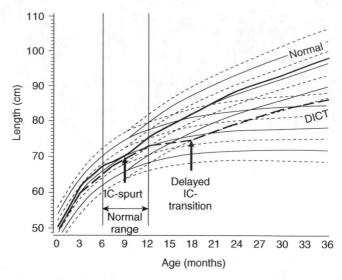

Fig. 4.E.2. The ICP model; mean, and ± 1, 2 and 3 SDS lines. The childhood growth stage sets in in Sweden between 6 and 12 months of age, and when delayed (DICT), it has a permanent effect on final adult height. Adapted with permission from Hochberg and Albertsson-Wikland (2008).

prepubertal stature (Liu, Jalil et al. 1998) and final adult height (Hochberg and Albertsson-Wikland 2008).

The negative impact of DICT on attained height is illustrated in Fig. 4.E.3 for childhood height and in Fig. 4.E.4 for adult height. The ICT age was defined for 1,720 Shanghai children, who were grouped by the ICT age to <9, 9–12, 12–15, and >15 months and followed until age 6 (Xu, Wang et al. 2002). The ICT gradient is apparent at age 6, when early ICT was associated with greater height. Each month's delay results in an adult height deficit of 0.15–0.2 standard deviation score (SDS) (Liu, Albertsson-Wikland et al. 2000; Hochberg and Albertsson-Wikland 2008), corresponding to a loss of 0.7–1.1 (mean 0.9) cm in final adult height. Hence, an early or late infancy–childhood growth transition within the normal range of 6–12 months accounts for an adult mean difference of 5.4 cm, and a severely delayed infancy–childhood growth transition to as late as 18 months may produce a deficit 10.8 cm, compared with individuals who have an early onset of the infancy–childhood growth transition at age 6 months.

Using a comparative cross-populations approach, evaluating natural fertility societies that live under different environmental conditions, we confirmed that the average weaning age of a society correlates strongly with the society average interbirth interval (Fig. 4.E.5). Much of the data we used relate to groups of hunters and gatherers because of their relevance to the vast majority of human evolutionary history. Interbirth intervals average 40 months when the first child survives, and less when the earlier child died. In traditional societies, some mothers wean after the next pregnancy is established and are both lactating and pregnant for a month, while very few have several months of periods between breast-feeding and pregnancy. The predicted energetic trade-off between the length of infancy and adult size was confirmed; the interbirth interval, as a surrogate measure of the infancy length, correlated negatively with adult bodyweight in both females and males. A similar

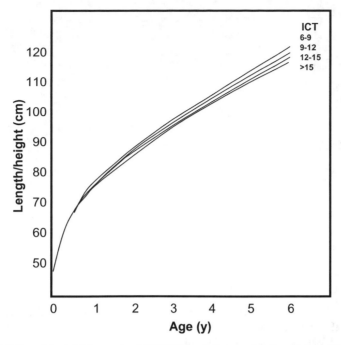

Fig. 4.E.3. Childhood height impact of DICT. The infancy–childhood transition (ICT) age was defined for 1,720 Shanghai children, who were grouped by the ICT age to <9, 9–12, 12–15, and >15 months and followed until age 6. The ICT gradient is apparent at age 6, when early ICT was associated with greater height. Data from Xu, Wang et al. (2002).

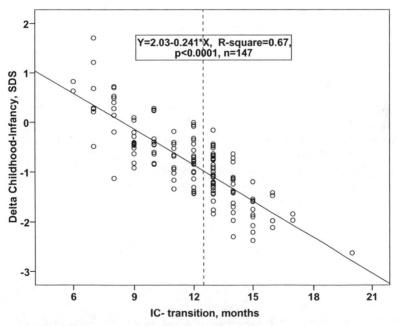

Fig. 4.E.4. Adult height impact of DICT. The length/height loss/gain (standard deviation score SDS units) from infancy to childhood as a function of age of the infancy–childhood transition. For each month of delay, a child lost 0.24 SDS. The difference between transitions at age 6 or 15 months amounts to as much as 2.1 SDS by the age of 2 years. Unpublished data by Hochberg, Jonsson, and Albertsson-Wikland.

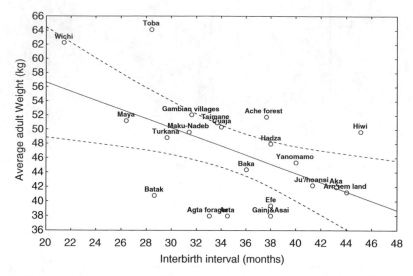

Fig 4.E.5. The interbirth interval. Adult body size of women in natural fertility societies as a function of the interbirth interval. Interbirth interval is used here as a surrogate to the weaning age (defining infancy) in the analysis of 21 natural fertility societies. Regression ($r = -0.538, p = 0.012$) and 95% confidence limits. Data from Gawlik et al. 2011.

relationship between weaning age and maternal bodyweight ($r = 0.91$) was previously reported for primates (Harvey and Clutton-Brock 1985). However, these authors proposed an opposite perspective of cause and effect, claiming that weaning age decreased due to the inability of a mother's metabolism to support the energetic needs of her infant that quadrupled its birth weight (Lee, Majluf et al. 1991; Bowman and Lee 1995). We proposed that with diminishing availability of critical resources, such as energy and nutrients, a trade-off for their costly allocation results in investment in a longer duration of breast-feeding, such as is associated with the typical decelerating growth at infancy (Hochberg 2009).

There have been several empirical applications of optimality models, designed to determine whether the onset and termination of reproduction and the size of interbirth intervals actually maximize fitness. The results of those analyses have been mixed. Analysis of the relationship of interbirth intervals to the survival of the sibling pair among the Ju!'hoansi San bushmen (Jo'hansi) of the Kalahari Desert showed that the 48-month interbirth intervals they usually practice were, in fact, optimal (Blurton Jones 1987). In contrast, similar analyses among the Ache of Paraguay showed that observed interbirth intervals were longer than optimal (Hill and Hurtado 1996).

In that respect, I find it interesting quote again from Darwin's *The Descent of Man*:

There is reason to suspect, as Malthus has remarked, that the reproductive power is actually less in barbarous, than in civilised races. We know nothing positively on this head, for with savages no census has been taken; but from the concurrent testimony of missionaries, and of others who have long resided with such people, it appears that their families are usually small and large ones rare. This may be partly accounted for, as it is believed, by the women suckling their infants during a long time; but it is highly probable that savages, who often suffer much hardship, and who do not obtain so much nutritious food as civilised men, would be actually less prolific. I have shewn in a former work ('Variation of Animals and Plants under Domestication'), that all our domesticated quadrupeds and birds, and all our cultivated plants, are more fertile than the corresponding species in a state of nature. It is no valid objection to this conclusion that animals suddenly supplied with an excess of food, or when grown very fat; and that most plants on sudden removal from very poor to very rich soil, are rendered more or less sterile. We might, therefore, expect that civilised men, who in one sense are highly domesticated, would be more prolific than wild men. It is also probable that the increased fertility of civilised nations would become, as with our domestic animals, an inherited character: it is at least known that with mankind a tendency to produce twins runs in families. (Mr. Sedgwick, 'British and Foreign Medico-Chirurgical Review,' 1863). Notwithstanding that savages appear to be less prolific than civilised people, they would no doubt rapidly increase if their numbers were not by some means rigidly kept down. The Santali, or hill-tribes of India, have recently afforded a good illustration of this fact; for, as shewn by Mr. Hunter (60. 'The Annals of Rural Bengal,' by W.W. Hunter, 1868, p. 259.), they have increased at an extraordinary rate since vaccination has been introduced, other pestilences mitigated, and war sternly repressed. This increase, however, would not have been possible had not these rude people spread into the adjoining districts, and worked for hire.

Phenotypic covariation, however, poses a major problem in those analyses. If healthier women have shorter interbirth intervals than less healthy women because they have larger effective energy resources, the estimated effect of interbirth intervals on adult size would be downwardly biased. In fact, studies examining natural variation among nonhuman organisms often show a positive, rather than a negative, relationship between fertility rate and offspring survival for the same reason. When fertility rate is experimentally manipulated among those same organisms, the relationship is reversed, as would be expected by a quantity–quality trade-off. Thus, Gambian women who had higher hemoglobin levels following the birth of a child exhibited both shorter interbirth intervals and higher child survival (Sear, Mace et al. 2001). It seems that women's physiology tracks its own condition in such a way as to maximize their individual fitness.

A delayed infancy–childhood growth transition is the main mechanism resulting in short stature in children living in poor areas of developing countries. In a community-based longitudinal study in Lahore, Pakistan, the median ages of the infancy–childhood growth transitions were 15, 13, and 10 months in suburban, village, and urban children, respectively, as compared to an average 9 months in the Swedish control group (Liu, Jalil et al. 1998). Among the poorest suburban children of Lahore, Pakistan, who suffer frequent infections, undernutrition and other afflictions of poverty, 35% have DICT, and among the poor children of Malawi, as many as 60% have a delayed infancy–childhood growth transition, and the average transition age is 2.4 years (Hochberg et al. unpublished). Many of them have an infancy–childhood growth transition that occurs at 3–4 years of age, which compromises their prepubertal height by as much as 15–20 cm, culminating in mean adult male and female heights of 162.5 and 155 cm, respectively (Zverev and Chisi 2004). An adult male height of 162.5 cm, and a back-calculated ICT at 3 years of age was also the weaning age of infancy and the average adult height of men in Italy before the commencement of the current secular trend in height in the mid-1800s (Fig. 2.C.1).

The gradient of age at the infancy–childhood growth transition among Sweden, Israel, China, Pakistan, and Malawi may represent a corresponding gradient of the quality of life provided to children of these countries.

The children of Tibet grow up under a unique high-altitude, nutritional, and traditional environment. A study of their growth during the initial 36 months of life gave a mean height and weight SDS of –1.53 and –1.05, respectively, in comparison with World Health Organization (WHO) standards (Dang, Yan et al. 2004). It is during the age range of 6–9 months (when affluent children make their transition to childhood growth) that the Tibetans' growth lingers behind. The prevalence of malnutrition in these children was 39% for severe short stature, 24% for underweight, and 6% for wasting, respectively, and rural children hung back more frequently than urban children. Slow growth during the transition period was clearly associated with altitudes.

Western medical literature describes the hanging back of growth of the children from developing countries as "stunting," with an implication of a pathological process. I deliberately used other phrases to denote this slow growth, as I view their stunting to be adaptive rather than pathological. Life in the altitudes of Tibet and the devastation of Malawi requires adaptation to a low-energy environment, and short stature is an adequate adaptive adjustment. The age when such adaptive short stature ensues is the ICT.

Nine fossils of adult *Homo erectus* subjects (1.8–1.3 million years before present) suggested a mean adult height of 162 cm (Leonard and Robertson 1997). Using the paradigm presented here, of 0.9 cm loss for every month of delay beyond the Swedish average of age 9 months at the ICT, and the mean Swedish conscripts' height of 1.79 m, *Homo erectus* ICT took place at 28 months, much like today's children of Malawi. Thus, *Homo erectus'* growth transition came about some 20 months before his apparent weaning age of 4 years, much like the 20 months between growth transition of Swedes and *Homo sapiens* life-history infancy at 30–36 months.

The ICP model proposed a period of overlap, when both the infancy and childhood growth components are operational, pointing to an inherent difficulty in the concept presented above. The childhood growth stage sets in 18–30 months before the childhood life-history stage, as defined by weaning in traditional societies at 2–3 years of age. Indeed, weaning takes place earlier in Western societies than the ethnographical weaning age. As we have no historical account of the infancy–childhood growth transition in previous years, it can only be speculated that in comparison to the transition age in a developing country, transition occurred later in ancient societies. Unfortunately, detailed infantile growth data are not available even for the past 150 years, yet, the conjecture it makes is that the secular upward trend of adult height over this period might have been due to changing the age of the infancy–childhood growth transition. Figure 2.C.2 shows the secular trend for Italians between 1850 and 1970. The mean difference of 13 cm would account for a shift of 13 months in the infancy–childhood growth transition age—quite plausible in view of the above discussion.

For the infancy–childhood growth transition to occur, the child must have a positive energy balance. The age when the infancy–childhood growth transition occurs is influenced and delayed by disease, when energy consumption increases rapidly (Karlberg, Jalil et al. 1994), and by undernutrition, gastrointestinal infection, and socioeconomic impediments, with insufficient energy supplies (Liu, Jalil et al. 1998; Liu, Albertsson-Wikland et al. 2000).

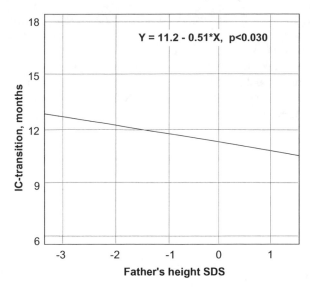

Fig. 4.E.6. Fathers' impact on the ICT. Negative correlation between fathers' height and infancy–childhood transition age. The transition age is earlier in infants of tall fathers, who presumably also had early transition) as compared to infants of short fathers. Mothers' heights do not correlate with the transition age, suggesting a transgeneration transmission of the transition age on the paternal side. Unpublished data by Hochberg, Jonsson, and Albertsson-Wikland.

There seems to be sufficient indirect evidence to propose that a delay in the infancy–childhood growth transition is a predictive strategy in adapting to an energy crisis during infancy. However, the contemporary crisis of Western societies is energy oversaturation rather than shortage, and the transition from infancy to childhood is also the age of shifting dietary saturation in affluent societies. It is suggested that in children with DICT, mismatching of the predictive adaptive response and other overfeeding results in the unwanted permutation of obesity with short stature (Hermanussen, Garcia et al. 2006). This is well documented for migrants from developing to richer countries, who are shorter but at higher risk for overweight and obesity (Varela-Silva, Frisancho et al. 2007). In a study of migrant Mayan-Americans, 11.5% of all children age 6–12 years were stunted in growth, and 48.6% were overweight, including 42.2% obese (Varela-Silva, Frisancho et al. 2007). Stunted children (as shown previously, 45% of whom have DICT), show higher susceptibility to the effects of a high-fat diet (Sawaya, Martins et al. 2004).

The ICT is also the mechanism for children to "channel into their genetic percentiles" as they transition from infancy to childhood growth stages. Indeed, a negative correlation was demonstrated between the age of the infancy–childhood growth transition and the heights of fathers, but not mothers' (Fig. 4.E.6), suggesting that the transgenerational transmission of transition age may be paternally derived.

Unlike the large volume of experimental data available for intrauterine programming/predictive adaptive responses, the infancy–childhood growth transition cannot be investigated in experimental animals; as mentioned, *Homo sapiens* is the only remaining hominine to have a life-history stage of childhood. Further investigation on this topic will have to rely on clinical research.

Unlike short periods of stunting at any other age, a delayed infancy–childhood growth transition is not followed by catch-up growth. I propose that the long transition period from infancy to childhood growth stages corresponds to a period of plasticity in growth that has evolved to adjust an individual's growth to environmental circumstances, such as limited energy resources. The ability of the individual to modify phenotypic response to altered environmental conditions is an evolutionary strategy and the basis for this adaptability.

F. WEANING FROM BREAST-FEEDING

And the mothers should suckle their children for two whole years for him who desires to make complete the time of suckling.

—*The Koran*, The Cow [2.233]

Intrauterine and infantile nutrition play a major role in lifetime health. They program, among other things, future growth and body composition. Feeding practices are fundamental elements of nutrition and are influenced by many factors, including personal and familial habits, maternal education, socioeconomic status, and cultural environment.

If we consider the life span of humans, the period of infancy, as defined by breast-feeding, is markedly shorter than, for example, the 60-month breast-feeding of the chimpanzee, which lives for 25 years in the wild. Weaning is followed in other social mammals by juvenility—a stage of independence for provision and protection; it is at about 30–36 months of age in human natural fertility societies (20 months for the Batak from the highlands of North Sumatra, and 42 months for the Ju!'hoansi San bushmen of the Kalahari Desert), and the definition of infancy and transition to childhood has been linked to weaning from breast-feeding. Breast-feeding controls ovulation through bursts of prolactin into the mother's body; the mother's return of ovulation is suppressed until lactation becomes less frequent. This sequence of events was called "the baby in the driving seat"(Lunn 1992), which envisioned babies as having taken control of their mother's hormonal balance while they needed the supply of breast milk to maximize their survival throughout infancy.

Trends of breast-feeding in developed countries changed, and currently mothers breast-feed, if any, from as short as 3 months to 2 years, well into the childhood growth stage. The American Academy of Pediatrics (AAP 2000) and the European Society for Pediatric Gastroenterology, Hepatology, and Nutrition (ESPGHAN) Committee on Nutrition recommended exclusive breast-feeding for the first 6 months of life (Agostoni, Braegger et al. 2009). The ability of breast milk to meet the requirements for macronutrients and micronutrients becomes limited with the increasing age of the infant. The WHO recommended continued breast-feeding for at least 2 years, and the AAP recommended it for at least 1 year (Gartner, Morton et al. 2005). For countries with a low infectious disease burden, the optimal duration with respect to health outcomes of any breast-feeding after the introduction of complementary feeding was concluded as uncertain for lack of data (Agostoni, Braegger et al. 2009).

Children who are exclusively or predominantly breast-fed for the first 4 months of life have a different growth pattern as compared to children who consume alternative energy sources in the first 4 months of life (WHO 1995). Whereas weaning implies transition to other foods than milk in traditional societies, in industrial societies mostly, infants will receive milk-imitating formulas. However, the imitation is far from perfect. Infants who followed the WHO recommendations for prolonged and exclusive breast-feeding, and who lived under conditions favoring the achievement of genetic growth potentials, show a decrease of growth progression in the first year and lesser obesity as compared with predominantly formula-fed infants (Hamill, Drizd et al. 1979). This observation of greater growth with early weaning has also been noted in beef calves (Meyer, Kerley et al. 2005).

As mentioned, weaning defines the end of infancy, yet weaning is a matter of maternal decision. Prolonged breast-feeding, as is the custom in areas of malnutrition in underdeveloped countries, has been perceived as a risk factor for impaired growth. The predictive adaptive concept of DICT suggests that during an energy crisis, it might be an appropriate trade-off to prolong infancy, defer the transition to childhood, stunt growth for a while, and end up adaptively shorter. This was cleverly tested in a rural area of Senegal by assessing nutritional status prior to weaning, and comparing it to the age at weaning (Simondon and Simondon 1998). These investigators found that length-for-age and weight-for-age during infancy were correlated with the duration of breast-feeding: Mothers deliberately nursed their infants longer (by definition prolonging infancy) when the infant's length or weight was smaller. The relationships remained at the same significant levels after adjustment for season of birth, mother's age, parity, height, occupation, and education. They concluded that the duration of breast-feeding is not determined by the disposition of a given mother only. The same mothers prolonged breast-feeding for undernourished children and reduced its duration for well-nourished children. While the direct reasoning and rationalization might be that mothers are aware of the mortality risk following weaning in a malnourished infant, this delay serves long-term fitness by producing adapted smaller adults in these underprivileged circumstances.

Modern women in affluent societies who breast-feed their infants wean them gradually around the age of 1 year. Later, we will discuss the energy consideration of transition from one life stage to the next; the larger babies of today may be ready to transition into childhood earlier than our ancestors did. By every definition of social, endocrine, or growth parameters of childhood, it seems to start around age 9–12 months, but it may be too early to come up with a new definition for modern infancy.

The infancy–childhood growth transition represents the age at which growth hormone begins to regulate growth significantly and reflects the control of growth by the growth hormone–insulin-like growth factor-I endocrine axis and target cell responsiveness to these hormones (Karlberg and Albertsson-Wikland 1988; Wit and van Unen 1992). The infancy–childhood growth transition occurs in parallel with a rise in serum levels of insulin-like growth factor-I and the insulin-like growth factor binding protein-3 during the second half of the first year of life. Children with a delayed infancy–childhood growth transition show a delay in the 6- to 12-month rise of insulin-like growth factor-I levels (Wang and Chard 1992; Leger, Oury et al. 1996) (Fig. 4.E.7). Moreover, an infancy–childhood growth transition is absent in children with growth hormone deficiency who receive no hormonal therapy (Karlberg and Albertsson-Wikland 1988).

Similar to human adaptive height control, it was shown that the underfeeding of a mouse during the early postnatal period delayed growth, whereas overfeeding

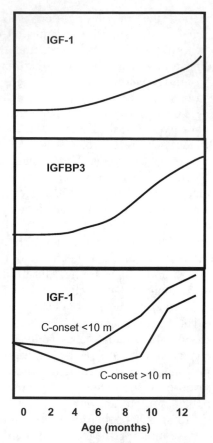

Fig. 4.E.7. Insulin-like growth factors and the infancy–childhood transition. The infancy–childhood growth transition occurs with the rise in serum levels of growth hormone-dependent insulin-like growth factor-I (upper panel) and insulin-like growth factor binding protein-3 (center) during the second half of the first year of life. Children with a delayed growth transition to childhood (lower panel) show a delay in the rise of insulin-like growth factor-I levels. Data from Karlberg and Albertsson-Wikland (1988); Wang and Chard (1992); Leger et al. (1996).

accelerated it (Kappeler, De Magalhaes Filho et al. 2009). In both cases, final body size was permanently altered. These investigators found coordinated alterations in pituitary growth hormone, plasma insulin-like growth factor-I and acid labile subunit (a growth hormone-dependent binder of insulin-like growth factor-I), and gene expression of hypothalamic growth hormone-releasing hormone GHRH during postnatal development. These changes were consistent with the observed growth phenotypes. They concluded that the insulin-like growth factor-I may play a role in modulating hypothalamic stimulation of the developing growth hormone system (Kappeler, De Magalhaes Filho et al. 2009).

It remains unclear how weaning may signal for the ICT, but indirect evidence may provide some clues. Weaning is marked by an adjustment in fatty acid metabolism, which is directly associated with the change in diet. The transition period characteristically reveals a decrease in plasma fatty acids and ketones, a decrease in fatty acid oxidation, ketogenesis, gluconeogenesis, and a rise in fatty acid synthesis (Little and Hahn 1990). In terms of the transition into the growth hormone-dependent childhood growth, the adipose tissue exerts a regulatory effect on the growth hormone–insulin-like growth factor-I axis mostly by modifying free fatty acids and the adipose tissue hormone leptin. Free fatty acids and growth hormone integrate a classical feedback loop; a rise in the former blocks the latter's secretion. This action is rapid, dose-related and is exerted at the pituitary level with some evidence for hypothalamic participation. A pharmacological reduction in free fatty acids enhances the secretion of growth hormone and eliminates the growth hormone blockade known to ensue in obesity and Cushing's syndrome (excess glucocorticoids).

Weaning presents also a transition from a high-fat to a high-carbohydrate diet. During infancy the baby feeds on a diet that is mother's-milk-based, relatively high in protein, and low in carbohydrate, which is characterized by a greater dependency on hepatic gluconeogenesis as the source of glucose for brain growth, development, and function. Gradual weaning and supplementation of grains-based foods, which ultimately increase the proportion of carbohydrate to protein and fat in the child's diet, require the emergence of an increased sensitivity to insulin action. This supports the utilization of carbohydrates and sugars, to replace milk's proteins and fats as metabolic fuels. The hormonal and metabolic changes that occur in response to the increase in plasma levels of glucose and insulin result in adaptation of digestive hormones and function. The parallel change in the nature and solidity of the ingested foods also affects the rate of enzymes and hormones maturation and their activity (Girard, Issad et al. 1993). Thus, to a large degree gastrointestinal maturation is driven by the foods ingested by the infant and child.

Indeed, the weaning process is marked by alteration of gut integrity, characterized by shortened villous length, disturbed absorptive/secretory electrolyte and fluid balances, increased mucosal permeability, decreased enzymatic activities, stimulation of proinflammatory cytokine gene expression, activation of heat shock proteins in the mucosa, as well as lowered level of mucins and goblet cell density. A 1- to 2-week adaptive stage to a solid diet based on plant ingredients is then observed (Montagne, Boudry et al. 2007). Breast-fed infants have smaller villi and crypts than bottle-fed infants, suggesting that crypt binary fission (duplication), associated with cylindrical growth of the small intestine, may be the predominant mechanism of epithelial growth during milk-feeding, whereas crypt hyperplasia predominates later during weaning (Cummins and Thompson 2002). Presumably growth factors in breast milk or those produced endogenously, such as corticosteroids, thyroxin, growth hormone, glucagon-like polypeptide-2, insulin, insulin-like growth factor-I, epidermal growth factor, erythropoietin, fibroblast basic growth factor, hepatocyte growth factor, keratinocyte growth factor, prostaglandin E2 polyamines, and transforming growth factor beta interact with specific mucosal growth factor receptors in the neonatal gut. It has been suggested that breast milk may also contain specific inhibitory growth factors for crypt hyperplasia.

Weaning is also associated with specific endocrine changes that may signal to the ICT. Breast-feeding influences circulating levels of leptin, ghrelin, and insulin-like growth factor-I in infancy, mainly during the first 4 months of life. During the first period, formula-fed infants compared to breast-fed show higher serum ghrelin, higher insulin-like growth factor-I and lower leptin concentrations (Savino, Fissore et al. 2005).

Human milk itself is a source of various hormones and growth factors that are obviously missing in any formula. Some of the hormones (insulin and adiponectin) may have a role in the maturation of the gastrointestinal tract, while others (leptin, adiponectin, ghrelin, resistin, and obestatin) are involved in food intake regulation and energy balance. The small intestine hormone motilin regulates interdigestive gastrointestinal contraction; its concentrations in lactating women's blood and milk are elevated for a period of at least 6 weeks (Liu, Qiao et al. 2004).

Adiponectin found in human milk has a role in inflammation, insulin sensitivity, and fatty acid metabolism. Milk adiponectin concentrations decrease throughout the lactation/infancy stage, whereby each month of lactation is followed by a decrease of 5–10% in milk adiponectin concentration (Woo, Guerrero et al. 2009). Milk adiponectin is associated with lower infant adiposity by 6 months of age.

In the developing rat, gastric density of ghrelin-generating cells is low on suckling and greatly expands after weaning, and similar observations have been made in humans (Bjorkqvist, Dornonville de la Cour et al. 2002). Gastric ghrelin expression and plasma ghrelin concentration are maintained at a lower level by delayed weaning (Fak, Friis-Hansen et al. 2007). Moreover, the relation between gastric ghrelin expression and body weight was altered by delayed weaning; timely weaned rats displayed a positive correlation between ghrelin expression and body weight, while no such correlation was evident in animals with delayed weaning. It seems that delayed weaning exerts a negative influence on the expression of ghrelin, the appetite hormone, and that commencement of solid food intake may trigger normal ghrelin expression. Thus, the growth hormone secretagogue ghrelin may constitute a hormonal link between weaning, appetite, the growth hormone–insulin-like growth factors axis and growth.

Comparison of ghrelin concentrations in lactating women showed lower values in late pregnancy "beesting" milk colostrum (mean 70 pg/ml), transitional milk (84 pg/ml) and mature milk (97 pg/ml) than in corresponding plasma samples (1st day 95 pg/ml, 10th day 111 pg/ml, and 15th day 135 pg/ml) (Woo, Guerrero et al. 2009). In contrast, the ghrelin gene product obestatin shows higher concentration in colostrum (539 pg/ml) and mature milk (528 pg/ml) than the corresponding blood levels of the mother (270 and 289 pg/ml, respectively). Leptin levels in colostrum (2 ng/ml) and mature milk (2 ng/ml) are more than fivefold higher than the corresponding blood levels (12 ng/ml).

Breast milk contains insulin-like growth factor-I, insulin-like growth factor-II, insulin-like growth factor binding proteins-1, -2, and -3. Concentrations of the latter are high from day 4 to day 6 and then decrease by days 10–12. In contrast, insulin-like growth factors-I and -II, as well as insulin-like growth factor binding proteins-1 and -2 show little change over the first 2 weeks after birth. Subsequently, all the insulin-like growth factor components show a moderate decline over approximately the first 1–3 months and then stable concentrations up until the transition to childhood at 9 months, when levels increase substantially (Milsom, Blum et al. 2008).

Other important growth factors as vascular endothelial growth factor, hepatic growth factor, and epidermal growth factor have been identified in breast milk especially in early breast milk (Kobata, Tsukahara et al. 2008).

5

CHILDHOOD

The childhood stage is peculiar to humans and has been defined by the stabilization of the growth rate, immature dentition, and weaning (while continuing to depend on older people for food), and on behavioral characteristics, including immature motor control (Bogin 1999a). The human childhood stage is among the cornerstones of humans' enormous evolutionary fitness. As a consequence of the childhood stage, more humans survive to adulthood than any other mammal. In addition to bipedalism, large brains and heads, and a spoken language, I view childhood among the foundations that define us as humans.

This period between infancy and juvenility is marked by sex hormones quiescence. The levels of testosterone in the boy and estrogen in the girl are extremely low, as is the concentration of the child's gonadotropins. During this stage, the hypothalamic–pituitary–gonadal axis is extremely sensitive to negative feedback by estrogen, thereby suppressing sex steroid levels to their lowest levels. In this quiescence of reproductive hormones, growth is faster and more stable as compared to that during the following juvenility period, when adrenal androgens appear and adiposity rebounds, consuming energy.

A. THE WEANLING'S DILEMMA

The vulnerability produced by humans' premature birth to fulfill the need to defer skull growth to the postnatal period has produced the "the weanling's dilemma" (Kennedy 2005): *Homo sapiens* have a longer period of dependency than other hominids, yet are weaned from breast-feeding earlier than any ape. The life stage

Evo-Devo of Child Growth: Treatise on Child Growth and Human Evolution, First Edition.
Ze'ev Hochberg.
© 2012 Wiley-Blackwell. Published 2012 by John Wiley & Sons, Inc.

of childhood emerges when the baby weans from breast-feeding, which for humans takes place at about 30–36 months of age, according to ethnographic observations in traditional societies and historical accounts. If we consider the extended life span of humans, the stage of infancy for *Homo sapiens* is markedly shorter than, for example, the 60-month infancy of the chimpanzee, which lives in the wild for about 25 years to have a mean adult life span after sexual maturity at age 10 of about 15 years (Hill, Boesch et al. 2001) (Fig. 1.D.2).

The evolutionary trade-off for the "weanling's dilemma" has been the introduction of an additional life history stage—childhood. Early weaning created an obvious hazard for the child, and even more so ever since hominids adopted carcasses as a new and valuable food approximately 2.6 million years ago, getting us into a dangerous competition with other carnivores. Yet many investigators suppose that the increased acquisition of energy-rich meat is among the most important evolutionary events that initiated the origin and success of the genus *Homo*. The trade-off for the new hazard was intellectual development, which became the primary focus of hominids' evolutionary selection.

The great lapse took place some 500,000 years ago with *Homo erectus*, whose transition from infancy to childhood took place before the first permanent molar tooth erupted. The new childhood stage was "deducted" from infancy; it reduced its length, and because of this gainful step, *H. erectus* enjoyed greater evolutionary fitness than any previous hominid. His populations increased in size and began to spread throughout Africa and other regions of the Old World.

The evolutionary gain of childhood lay in the mother's freedom to cease breast-feeding her 3-year-old infant in order to become pregnant again. This enhancement of reproductive output required the social interaction of an extended family and tribal ties, and as is evident from the consequences of this evolutionary leap, it did not put the mother, her infant, or her older children at risk. The Igbo and Yoruba tribes of Nigeria say that "it takes a village to raise a child."

Once the infant is weaned, the child is no longer solely dependent on his mother for provision. The entire family, and sometimes the entire tribe, share responsibility of provision and protection. Puppies and little kids look cute to most adults in all cultures, and cuteness elicits loving care by adults. Evolution has made these positive emotions a crucial element in the childhood life-history stage. In a study on the role of the extended family in a rural area of the Gambia, it was demonstrated that having a living mother, maternal grandmother, or elder sisters had a significant positive effect on the survival probabilities of children, whereas having a living father, paternal grandmother, grandfathers, or elder brothers had no such effect (Sear, Steele et al. 2002). Family and tribal care is a unique hominid adaptation, for no other primate or mammal (elephants may be one such exception) is so actively involved in feeding the broader family's young (Lancaster and Lancaster 1983). A father role is a much later development in human sociology. It was not until the post-Ice Age extinction of many wild animals and the transition to shepherds families that the female primary groups accepted men as being "the father" to provide maximum protection for his wife and children.

The long period of food provisioning and protection, extending from 3 to 6 years, largely defines the childhood stage of human life history. By the age of 7, the four permanent molars have usually erupted and the permanent incisors have begun to replace "milk" incisors (Fig. 4.B.4).

Toward the end of this stage, children become able to eat the same foods as adults, new learning and behavioral capabilities enable greater social independence, the gait becomes adult-like (Bogin 1999a), and cognitive and emotional developments permit new levels of self-sufficiency. Seven-year-olds can perform many basic tasks, including food preparation, infant care, and other domestic tasks with little or no supervision.

Jean Piaget defined childhood as the cognitive preoperational stage (Piaget 1936) (Table 4.1). The child increasingly uses verbal representation, but speech is egocentric. Applying his new knowledge of language, the child begins to use symbols to represent objects. Early in this stage he also personifies objects. He is now better able to think about things and events that are not immediately present. He is oriented to the present, but still has difficulty conceptualizing time. His thinking is influenced by fantasy—the way he'd like things to be—and he assumes that others see situations from his viewpoint. He takes in information and then changes it in his mind to fit his ideas.

In summary, the evolutionary advantages of childhood are apparent and contributed to the reproductive fitness of those hominids who acquired it: (1) By reducing the length of infancy and adding childhood, hominid mothers gained greater lifetime fertility than any ape, and (2) as a consequence of the family-oriented childhood stage, more hominids who acquired childhood survived to adulthood than any other mammal. These fitness arguments increased as childhood has prolonged from 1–2 years in the *Homo habilis* to 3–4 years in *Homo sapiens sapiens.*

B. THE GRANDMOTHER THEORY

Observing nature in the wild gives the impression that animals are mostly selfishly hostile to each other. The truth is that cooperation or cooperative behaviors are common, and turn out to be beneficial to other members of the same species. We have all observed the cooperation of ants and bees in their nests, but these are small and distant creatures. Tribal life is an important case in point of a cooperation model. The theory of "kin selection" refers to apparent strategies in evolution that favor the reproductive success of an organism's relatives, even at a cost to their own survival and/or reproduction (Hamilton 1963; Smith 1964). Genes for such cooperative operations are conserved because they help preserve their owners from extinction. The continuity of his genes is the only reward the altruistic individual receives.

That grandmothers behave altruistically is a cross-cultural social tenet. Otherwise, why does a human's life history include two decades beyond reproduction? But women as young as their mid-30s can be grandmothers, and we need to be reminded that it is very common that women are both mothers of their younger children and grandmothers relative to the offspring of their oldest children at the same time. In order to contribute to the well-being of both generations of children, this woman needs to be in residential contact with the children and able to perform work, especially food collection but also the tasks of the village such as water collection, hut construction, childcare, food processing and cooking, and so on to meet the needs of her own and her adult child's household. Despite some objections, it is now accepted that a postmenopausal life stage is evolution-based, which is suggested by the fact that the decline of ancestral age-specific fertility persisted in our genus, while senescence in other aspects of physiological performance slowed down (Hawkes 2003).

The grandmother theory is an attempt to explain why menopause, rare in mammal species, arose in human evolution and how a long post-fertile period up to one-third of a female's life span could confer an evolutionary advantage (Williams 1957). The theory maintains that this unusual feature of human demography has evolved because of the contributions that the postmenopausal female makes toward the fitness of her children and grandchildren. She can monitor the health and well-being of her child and grandchildren, and contribute to their well-being, and she turns out to be the best gatherer in the community. On a mere genetic basis, women retain greater evolutionary advantage by caring for their daughters, who carry 50% gene identity, and grandchildren (25% identity) than by investing in more of their own children or the risks of multipara birth. A study of the Hadza tribe, a small group of hunter-gatherers in Tanzania, showed that while caring for their daughters and grandchildren, women in their 50s–70s and beyond are also among the most diligent members of the group, gathering more food than almost any of their tribal peers (Blurton Jones, Hawkes et al. 2002). Proponents of the grandmother theory maintain that postmenopausal survival, like big brains and upright posture, is among the biological traits essential for classifying us as humans.

While the grandfather carries over the same 25% of his genes to his grandchildren, we have to distinguish between grandmothers and grandfathers. When women of the Ju!'hoansi San of the Kalahari Desert become widows, they are likely to concentrate all their efforts on assisting their adult children (and continuing to raise their own if any are still dependent), while men who are widowed are more likely to marry again and may be caught up in providing assistance to the new wife's adult children. No doubt these matters differ according to the cultural rules of the particular group, but they are likely to be important events in the lives of individuals from any hunter-gatherer group (Howell 2010).

It was claimed by others that most of the benefits to longevity of the grandmother derive from helping their offspring rather than their grandchildren—the so-called mother hypothesis. (Lancaster and Lancaster 1983). It was argued that rather than the grandmother being a determinant of our longevity, the reverse may be so—our longevity is a central determinant of the grandmother's existence: The length of a female's post-reproductive life span was reflected in the reproductive success of her offspring and the survival of her grandchildren (Lahdenpera, Lummaa et al. 2004).

C. GROWTH OF THE CHILD

The end of the rapid growth deceleration of infancy marks the beginning of childhood, when, quite uniquely to modern humans, the growth rate levels off to a quasilinear rate of an average of 7 cm per year with minimal if any acceleration or deceleration (Fig. 2.B.1). Sexual dimorphism is almost nonexistent in this stage of quiescent hypothalamic–pituitary–gonadal axis in the boy, and semi-quiescent in the girl (Klein, Baron et al. 1994). Despite 10-fold higher estrogen levels, girls grow as fast as boys during childhood, with somewhat of an acceleration at the end of this phase [the so-called mid-childhood or mid-growth spurt (Molinari, Largo et al. 1980)], before the ensuing juvenility slows down growth.

The stable growth of children defers the total energy costs of rearing them by the family and tribe, and appears to enable parents to support multiple dependents of different ages at any one time (Dean 2007). Were our growth curve more like that of chimpanzees, which transit directly from infancy to juvenility and do not show a stable period of growth, then it would cost various modern hunter-gatherer mothers between 1,267 and 1,503 kcal/day more to support the same number of offspring (Gavan 1953).

Preliminary data suggest that the duration of this stable growth period may have a major impact on final height. Assuming equal growth and timely transition from infancy to childhood as well as during juvenility adolescence, each year of a childhood period adds to the child's final height as much as 6.2 cm in boys and 6.9 cm in girls. A shorter childhood and earlier transition to juvenility in girls give boys a small advantage in total childhood growth. The story of the Pygmy will be told later, but it can be mentioned here that the Pygmy have a shorter childhood stage, and they transition earlier into juvenility and adolescence with significant compromise of their adult height.

D. ENDOCRINE ASPECTS OF CHILDHOOD GROWTH

The driving force behind the transition from infancy to childhood growth in both boys and girls is the growth hormone–insulin-like growth factor-I axis. The infancy–childhood growth transition at a mean age of 9 months (also the age at which growth hormone becomes operational) reflects the control of growth by the growth hormone–insulin-like growth factor-I axis endocrine axis and their target cell responsiveness (Karlberg and Albertsson-Wikland 1988; Wit and van Unen 1992; Hochberg 2002a). The infancy–childhood growth transition occurs in parallel with a rise in serum levels of the growth hormone-dependent insulin-like growth factor-I and insulin-like growth factor-I binding protein-3 during the second half of the first year of life. Children with a delayed infancy–childhood growth transition show a delay in the 6- to 12-month rise of insulin-like growth factor-I levels (Wang and Chard 1992; Leger, Oury et al. 1996) (Fig. 4.E.7). Moreover, the growth transition is absent or very delayed in children with growth hormone deficiency who receive no hormonal therapy (Karlberg and Albertsson-Wikland 1988).

6

JUVENILITY

Other than humans, all other mammals (including the great apes) transit directly from infancy to juvenility without passing through the childhood stage (Fig. 1.D.2). Comparison with the African apes suggests that the timing of transition to juvenility, as measured by adrenarche in chimpanzees may be similar to that in humans, although the full course of age-related changes in dehydroepiandrosterone-sulfate (DHEAS) and their relationship to reproductive and brain maturation are not clear (Campbell 2006).

This chapter defines juvenility as a distinct clinical life-history stage, characterizes it in terms of its unique endocrine and body composition changes, relates these changes to social assignments and psychological maturation, describes the plasticity in the transition from childhood into juvenility and from juvenility to adolescence, and claims that this life-history stage is endowed with programming—a predictive adaptive response for a thrifty phenotype, metabolism, and body composition. In Chapter 2, Section E, I have shown that juvenility is also the programming phase for the fertility strategy based on the juvenile's parental attachment experience.

Initial classifications of life history defined pubarche as the onset of juvenility (Bogin 1999a). This treatise shows a much earlier age of transition to juvenility; thus, pubarche is a late event during juvenility and quite subjective, too. However, it is clinically obvious and studies of premature pubarche may shed light on the impact of early or late juvenility. The timing of transition from childhood to juvenility is associated with changes in growth and final adult height (see Chapter 12).

Evo-Devo of Child Growth: Treatise on Child Growth and Human Evolution, First Edition.
Ze'ev Hochberg.
© 2012 Wiley-Blackwell. Published 2012 by John Wiley & Sons, Inc.

A. THE SOCIAL/COGNITIVE DEFINITION OF JUVENILITY

As mentioned, the preceding childhood stage is exclusively a human innovation; it is defined by a stabilizing growth rate, by immature dentition and weaning while the child continues to depend on older people for provisions and protection, and by behavioral characteristics, including immature motor control. The juvenility stage offers opportunities to prepare for the social complexities of adolescence as well as adulthood.

The psychologist Sheldon White (1928–2005) called it "the five-to-seven-year shift" or "the age of reason and responsibility" (White 1996). This is the stage when the brain reaches its final size, and equipped with adult molars, primates move on to juvenility to forage independently for food and care for themselves. Whereas chimpanzees make the move directly from infancy, humans, who have a shorter infancy, initiate juvenility after a period of childhood.

In modern societies the transition to juvenility coincides with the age when children go to school, and compete to some extent with adults for food and space. They can now understand and model the roles of their parents and older children. As they join the adult social activities, they develop the typical late juvenile strong odor that repels the opposite sex to avoid sexual aggression (Weisfeld, Czilli et al. 2003). In a study that explored kin recognition through olfaction, these investigators found that mutual olfactory aversion occurred only in a father–daughter and a brother–sister nuclear family relationships. Recognition occurred between opposite-sex siblings but not same-sex siblings.

Cognitive and social advances accompany the physical changes induced by adrenarche. Developmental psychologists refer to this period as "middle childhood": a period of cognitively concrete operations, when children become less dependent on their parents for support and begin to interact with other adults and peers (Table 2.1). The juvenile builds up organized and logical thoughts, an ability to perform multiple classification tasks (order objects in a logical sequence), and comprehend the principle of conservation. Thinking becomes less transductive and less egocentric, and the child is capable of concrete problem solving. He acquires discipline in writing, reading, spelling, verbal memory, manual training, the practice of instrumental technique, proper names, and drawing drills in arithmetic. [G. Stanley Hall (1844–1924) quoted by White (White 1996)].

To become independent, the juvenile learns complex feeding skills: when and where to find food, and how to hunt with the group. In that respect, it is interesting that around age 6, a systematic process of brain gray-matter reduction occurs in the primary association areas (Gogtay, Giedd et al. 2004), which will be complete in the prefrontal cortex in the 20s. Gray-matter reduction represents synaptogenesis during this period (Paus 2005).

White suggested several reasons to propose that this psychological package is intrinsically neurobiological (White 1996):

(1) Electroencephalographic rhythms stabilize in a basically adult pattern.
(2) Blindness beginning before the shift results in an absence of visual memories, but later onset blindness preserves such memories.
(3) Loss of a limb before the shift leaves the child with no sensation of the missing limb, but later often produces a phantom limb.

(4)   There is an inflection point in brain size growth such that brain growth slows markedly after the shift.

(5)   Myelination is largely complete in most major circuits, producing effectively adult levels of axon conduction, especially in longer pathways.

(6)   There are changes in the granule cells of the dentate gyrus of the hippocampus, a structure crucial to the encoding of the associative system.

(7)   Proliferation of synapses peaks in certain cortical regions prior to normal regression.

(8)   Cortical energy consumption peaks as measured by PET scanning.

B. PALEOANTHROPOLOGICAL JUVENILITY AND TEETH ERUPTION

For the paleoanthropologist, the transition from childhood to juvenility is associated with the eruption of permanent molars (Fig. 4.B.4), for it is the fossil that they usually come upon. The age when permanent molars erupt in *Homo sapiens* is 6 years. A comparative study across 21 primate species found the age of the first molar eruption to be highly associated with humans' main evolutionary thrust—brain weight ($r = 0.98$) and a host of other life-history variables (Smith 1994). The strength of the correlations seemed best explained by the robust tooth development in response to environmental perturbations, especially when compared to such life-history variables as the age of sexual maturation or the first birth. Thus, the mean age of tooth eruption is a good overall measure of the maturation rate of a species, and valuable in the context of the present theme—the transition from childhood to juvenility. Interestingly, data for dental eruption in 1837 showed similar eruption ages to those known today (Saunders 1837); unlike the secular trend in the transitional age to childhood and adolescence (at a time of a marked worldwide upward trend for height), transition into juvenility, as assessed by teeth eruption, has not changed much over the past 170 years (Helm 1969).

The transition age to juvenility as determined by the eruption of the first molar may be even longer standing. A study using novel techniques to estimate the chronological age of an ancient *Homo sapiens* from Jebel Irhoud in Morocco,[1] dated these fossils to 160,000 years before the present (Smith, Tafforeau et al. 2007). They showed that the age of the lower incisor tooth, which erupts in contemporary children at 6–7 years of age, was much the same then as it is today, suggesting that in terms of teeth eruption, the transition age to juvenility has not changed throughout the ~200,000 years of modern humans.[2] Interestingly, the Neanderthals' permanent molars erupted at a comparable age of 6.5 years (Coppa, Manni et al. 2007).

C. ADRENARCHE

It is difficult to decide when to set the transition age to juvenility, as some parts of the juvenile package may start early, and others a year or two later. Should the

---

[1] The archaeological cave site near Sidi Moktar, about 100 km west of Marrakesh.
[2] "Modern humans" refers to *Homo sapiens* species as of 200,000 years before the present, of which the only extant subspecies is *Homo sapiens sapiens*, about 100,000 years before the present.

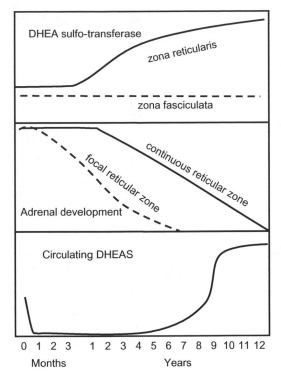

Fig. 6.C.1. Age-related changes in adrenal androgens and zona reticularis. Immunoreactivity of DHEA sulfo-transferase (SULT2A1) (upper panel) in the zona fasciculata (dashed line) and reticularis (solid line). Data by Suzuki et al. (2000). Age-dependent development of a focal (dashed line) and a continuous reticular zone (solid line; middle panel). Data by Grumbach and Styne (2003). Serum DHEAS concentrations in children (lower panel). Data by Korth-Schutz et al. (1976).

transition to juvenility be defined by adrenarche? For the onset of adrenal androgen generation, this was generally considered to be around age 7–8 for girls and 8–9 for boys. Yet a closer look at adrenal androgen levels suggests an earlier age (Figs. 6.C.1 and 6.C.2) (Korth-Schutz, Levine et al. 1976; Sizonenko 1978; Suzuki, Sasano et al. 2000): as early as ages 5–6. In preparation for androgen secretion, the adrenal zona reticularis emerges as early as age 3–4 initially as a focal reticular zone (Havelock, Auchus et al. 2004; Arlt, Martens et al. 2002; Auchus and Rainey 2004).

While human beings and chimpanzees exhibit adrenarche, other primates such as the baboon and rhesus monkey do not, and the adrenals of most other mammals produce little or no DHEA (Arlt, Martens et al. 2002). Thus, the acquisition of adrenarche is a very recent evolutionary event. For the DHEA-generating enzyme 17,20 lyase, the human and chimp enzymes differ at only two amino acids, whereas the human/chimp enzyme differed from the baboon or rhesus enzyme by 25–27 residues (95% identity) (Arlt, Martens et al. 2002).

DHEA secretion, like that of cortisol, is stimulated by the andrenocorticotrophic hormone (ACTH). Thus, after the transition to juvenility, but not during infancy and childhood, cortisol and DHEA respond in a similar way to stress and the environment.

100 JUVENILITY

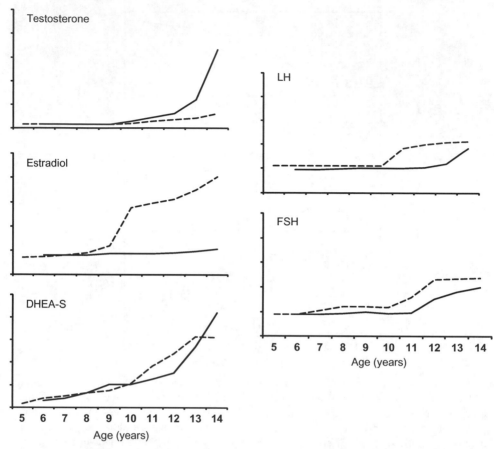

Fig. 6.C.2. Juvenility-related changes in sex hormones in boys (solid lines) and girls (dashed lines), relative levels without units.

The endocrine control of adrenarche has been the topic of several excellent review articles (Ibanez, Dimartino-Nardi et al. 2000; Auchus and Rainey 2004), and the following discussion focuses only on those aspects that may explain the evolutionary perspectives of juvenility. Serum DHEA and DHEAS rise progressively throughout juvenility (Palmert, Hayden et al. 2001), with effects on a wide variety of physiological systems, including neurological (Kroboth, Salek et al. 1999), immune (Chen and Parker 2004), and somatic growth and development (Zemel and Katz 1986; Arquitt, Stoecker et al. 1991).

It has been suggested that the primary effects of DHEA/DHEAS in humans is as a neurosteroid affecting neurological functions and modulating mood (Hunt, Gurnell et al. 2000; Suzuki, Wright et al. 2004). Thus, in parallel with adrenarche and the adrenal generation of DHEA and DHEAS, the child attains his social juvenility to function as boy hunter or a girl gatherer. In that respect, adrenarche is timely at the maturation age of the cerebral cortex, which extends from age 6 to 20. Campbell suggested three mechanisms by which DHEAS may promote changes in behavior and cognition (Campbell 2006), all of which are in line with the evolutionary significance of juvenility: (1) acting on the amygdale to reduce fearfulness and allow for the expression of an increased range of social interactions with unfamiliar individuals, as the juvenile independently cares for his new needs and interacts with peers; (2) acting on the hippocampus to promote memory, and social and cognitive capacity, as he joins in some adults activities; and (3) acting as an allosteric antagonist to the GABA receptor in memory improving and its antidepressant effects. In addition, DHEAS may play a role in synaptogenesis and cortical maturation, and may help wire the brain in response to existing social environments. It is conceivable, although it needs experimental evidence, that these brain effects of DHEAS are required to prepare the central nervous system for puberty, both in its psychosocial sense and in setting the scene for maturation of the hypothalamic–pituitary–gonadal (HPG) axis. During both juvenility and adolescence, DHEA is secreted regardless of the HPG axis (Palmert, Hayden et al. 2001), and acts independently of gonadal steroids in promoting sexual behaviors (Halpern, Udry et al. 1998).

It has been suggested that DHEA and DHEAS have a role in the early juvenile mid-childhood growth spurt, but sexual dimorphism in the growth spurt suggests that DHEA and DHEAS are not the only control mechanisms for childhood growth. In fact, as the weak androgenic DHEA levels rise during juvenility, growth decelerates. It was shown that DHEA suppressed bone growth by acting directly at the growth plate through the estrogen receptor. Such growth inhibition is mediated by decreased chondrocyte proliferation and hypertrophy/differentiation and by increased chondrocyte apoptosis (Sun, Zang et al. 2011).

The literature on adrenarche has been dominated by the view that it precedes any activity of the HPG axis. Yet, the HPG axis is operational during juvenility and shows unique changes in both sexes (Mitamura, Yano et al. 1999, 2000). Moreover, gonadal steroids influence the unique juvenile growth pattern. Evaluation of the diurnal rhythm of estrogens indicates that girls show a rise in early morning estradiol levels at least 2 years before the onset of clinical puberty, and in the middle of juvenility (Norjavaara, Ankarberg et al. 1996), and that premature ovarian failure and oophorectomy in young as well as postmenopausal subjects precipitate an earlier decline in DHEA levels (Cumming, Rebar et al. 1982). By the same token, a significant positive correlation was observed between DHEAS levels, body weight, and each stage of breast development before and after menarche (Murakami, Kawai et al. 1988). Indeed, estrogens are known to suppress adrenal cortex 3β-hydroxysteroid dehydrogenase activity (DHEA-degrading mechanism) and enhance 17,20-lyase activity (DHEA-generating mechanism), resulting in higher DHEA and DHEAS levels in response to estrogens (Miller 1999).

D. JUVENILE BODY COMPOSITION

The life histories of hunter-gatherers are remarkably similar to those of people living in modern societies. Hunter-gatherer children are largely supported by their parents, and after a period of decelerating net food production (food production vs. consumption) during childhood, net production levels off at transition to juvenility (Kaplan and Robson 2002).

Adiposity rebound is an important indicator for the transition into a juveniles' body composition (Fig. 6.D.1) (Hochberg 2008). The adiposity rebound corresponds to the second rise in the age-related body mass index (BMI) curve that occurs between the ages 4 and 6 years. Its sexual dimorphism corresponds very well to that observed in the second derivative growth curves, with the mean boys' rebound at age 68 months, as compared to girls with a mean adiposity rebound 6 months earlier (Fig. 6.D.2)

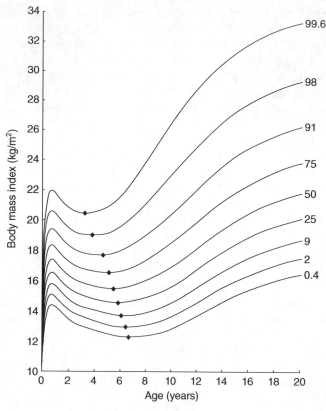

Fig. 6.D.1. Adiposity rebound. The age at adiposity rebound is inversely related to the body mass index percentiles, shown to the right of the parallel lines. The adiposity rebounds are marked with dots for each line, showing the younger age of adiposity rebound in the obese. Adapted with permission from the British 1990 girls BMI chart by Cole et al. (1995).

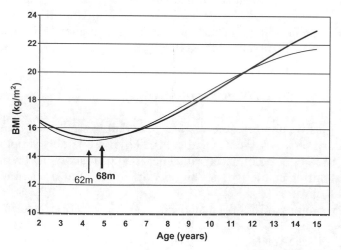

Fig. 6.D.2. Sexual dimorphism in adiposity rebound. Median BMI values for boys (thick line) and girls (thin line).

An early adiposity rebound is observed in overweight children and is associated with an increased risk of obesity at adulthood (Kerem, Guttmann et al. 2001), suggesting that the child's body habitus is programmed for early transition from childhood to juvenility (Rolland-Cachera, Deheeger et al. 2006). The typical pattern associated with an early adiposity rebound is a marked increase in BMI during juvenility that will exacerbate during adolescence. This pattern is recorded in children of recent generations as compared to those of previous generations, owing to the trend of a steeper increase of height as compared to weight in the first years of life.

Adiposity rebound may be the first clinical sign of juvenility, or it may be the signal that turns the transition on. It is interesting to note that even lean girls with precocious adrenarche have higher than control levels of insulin-like growth factor-I, insulin-like growth factor binding protein-3 and leptin (Guven, Cinaz et al. 2005), as mechanisms that may transmit the signal of energy readiness.

Two important clinical observations may be relevant to understand the importance of this age with respect to body composition. The first is a study of obese boys with a family history of obesity and metabolic syndrome that disclosed the transition to juvenility coincides with obesity onset (Kerem, Guttmann et al. 2001); these boys had an onset of obesity at a mean age of 6.4 years, as compared to 2.3 years in obese children without family history of the metabolic syndrome. Children with obesity onset at juvenility had a truncal (android) distribution of fat, and their fasting blood glucose was higher while HDL/total cholesterol ratio was lower.

The second observation is that it is around the age of 4–5 that children with Prader-Willi syndrome (PWS)[3] become progressively overweight, while developing the typical habitus of high body fat mass/low body muscle mass (Lindgren, Barkeling et al. 2000). Indeed, PWS patients tend to have premature pubarche, with higher DHEAS levels at adolescence, as compared with control subjects (Unanue, Bazaes et al. 2007), normalizing at adulthood (Hirsch, Eldar-Geva et al. 2009). Whereas the pathophysiological mechanisms that underlie obesity in PWS are poorly understood, suggestions of increased insulin sensitivity (Schuster, Osei et al. 1996) may imply a role for insulin in the link between obesity and juvenility.

The close proximity to adiposity rebound suggests a link of transition to juvenility to energy supply. In support of this, increases in the juvenility hormone DHEAS concentrations correlate positively with increases in BMI (Remer and Manz 2001). It has been suggested that as brain and somatic growth tapers off during juvenility, energy allocation that was formerly associated with brain and somatic growth is temporarily stored as abdominal fat, in order to support the energetically costly accelerating growth during the upcoming adolescence (Campbell 2006). Indeed, the transition age to juvenility is strongly linked to the age at onset of puberty; patients with precocious puberty had an early adrenarche (Palmert, Hayden et al. 2001), and those with delayed puberty or hypogonadotrophic hypogonadism had a late transition to juvenility (Van Dop, Burstein et al. 1987).

It is interesting to note that while adiposity accelerates in mass during juvenility, bone decelerates in relative mass and accruing minerals, while the rise in bone

---

[3] Prader-Willi syndrome is a contiguous gene syndrome resulting from deletion of the paternal copies of the imprinted SNRPN, the necdin gene, and possibly other genes within the chromosome region 15q11-q13. It is characterized by diminished fetal activity, obesity, muscular hypotonia, mental retardation, short stature, hypogonadotropic hypogonadism, and small hands and feet.

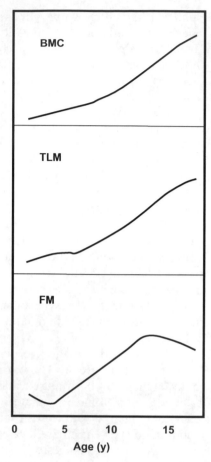

Fig. 6.D.3. Body composition changes at the transition to juvenility. Body composition as a function of age in boys: bone mineral content (BMC, upper panel), lean tissue mass (LTM, middle panel), and fat mass (FM, lower panel). Data from Ellis (1997); Ellis, Abrams et al. (1997).

mineral density reaches its nadir (van der Sluis 2002) (Fig. 6.D.3). These two tissues—bone and adipose—mature out of a common stem cell, which ultimately differentiates into osteocytes, chondrocytes, or adipocytes. The plasticity to differentiate either route is a feature of the stem cell, which responds to the environment by means of a complex array of hormones and growth factors. *In vitro* and *in vivo* studies strongly support an inverse relationship between the commitment of mesenchymal stem cells or stromal cells to the adipocyte and osteoblast lineage pathways (Gimble, Zvonic et al. 2006).

Juvenility and increasing adrenal androgens concentrations are associated with an increase in muscle mass and bone mineral content. The association of enhanced adrenal androgen generation in congenital adrenal hyperplasia with muscularity is well documented (Rodda, Jones et al. 1987). Accordingly, an increase in fat-free, lean body mass is evident around age 5, which is greater in girls than in boys, apparently as part of the female mid-childhood spurt (Fig. 6.D.3) (Ellis 1997; Ellis, Abrams et al. 1997; Teramoto, Otoki et al. 1999). As a consequence of increasing fat-free mass, bone remodeling accelerates; in a study of 205 healthy juveniles, a significant influence of muscularity on periosteum modeling was found, with positive correlation of adrenal androgens with cortical density and bone mineral content (Remer, Boye et al. 2003).

The adrenal reticularis is not the only gland that takes a turn with the transition to juvenility. The growth hormone–insulin-like growth factor axis activity is enhanced in parallel with the rise in adrenal androgens (Guercio, Rivarola et al. 2002; Guercio, Rivarola et al. 2003) and in girls more than in boys. These connections are further described as the predictive adaptive importance of juvenility onset is entertained.

E. GROWTH OF THE JUVENILE

After the childhood period of constant growth rate, and the "mid-childhood spurt" (greater and earlier for girls than for boys), a decline in the rate of growth signifies a transition into a new life stage, giving late juvenility the slowest growth rate since birth (Hochberg 2008). Growth becomes slowest when the child is ready to transition into adolescence and the concurrent pubertal growth spurt starts (Zemel and Katz 1986). This juvenile growth pattern is mostly evident from the first (velocity) and second derivative (acceleration) curves; these suggest a mean onset of juvenility at age 4.5 years for girls and 5.5 years for boys (Fig. 2.B.1). A trade-off for this deceleration (at a time when brain growth is almost complete) may have to do with the learning required for living within the social hierarchy of the group without posing a physical threat of a big body (Janson and van Schaik 1993).

Androgens in general enhance growth, as is well known from the adolescent growth spurt. Many of us were puzzled for years why it is that following the transition from childhood to juvenility through adrenarche, children decelerate their growth. A recent study provides evidence for a direct inhibitory effect of adrenal DHEA on the growth plate enchondral ossification process (Sun, Zang et al. 2011). In this study, DHEA suppressed metatarsal growth, growth plate chondrocyte proliferation, and hypertrophy/differentiation. In addition, DHEA increased the number of apoptotic chondrocytes in the growth plate. In cultured chondrocytes, DHEA reduced chondrocyte proliferation and induced apoptosis. To do so, DHEA is aromatized to estrone, which utilizes estrogen receptors and nuclear factor-κB DNA binding activity to slow growth.

Slowing of growth coincides with the social assignment of juveniles as they join the adults' society for hunting or domestic tasks. They are not to be perceived as a threat to adults, whom they expect to coach them in the art of being a true associate of the clan. The importance of maintaining slow growth during juvenility, while keeping reproductive hormones suppressed, is revealed when a child develops precocious puberty. Desensitization of the HPG axis to sex steroids' negative feedback is associated with an intense growth spurt that concludes with the fusion of the epiphyseal growth plates and cessation of growth.

Analysis of auxological (the science of growth) changes provides interesting insight into changing body proportions. Leg length as a function of age shows a clear acceleration in the relative lower limb growth as juvenility commences between the ages of 5 and 6, and slightly later in boys as compared with girls (Fig. 6.E.1); this is mostly evident in the second derivative acceleration curve, in accordance with the juvenile's new social role of independence for provisions and protection. Sitting height (Fig. 6.E.2) and the bi-iliac diameter (Fig. 6.E.3) also show acceleration,

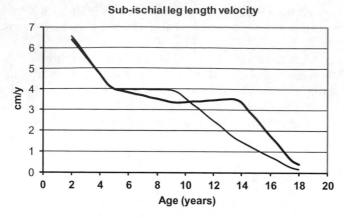

Fig. 6.E.1. Subischial leg length. Mean subischial leg growth velocity (the first derivative, cm/y). The first derivatives were calculated from Gerver and de Bruin (1996).

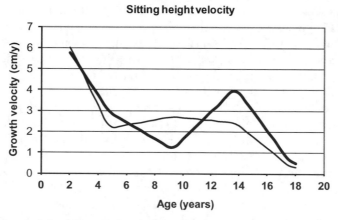

Fig. 6.E.2. Sitting height. Mean sitting height velocity (the first derivative, cm/y) for boys (thick line) and girls (thin line). The first derivatives were calculated from Gerver and de Bruin (1996).

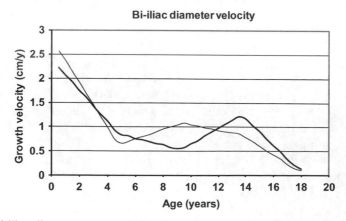

Fig. 6.E.3. Bi-iliac diameter. Mean bi-iliac diameter velocity (the first derivative, cm/y) for boys (thick line) and girls (thin line). The first derivatives were calculated from Gerver and de Bruin (1996).

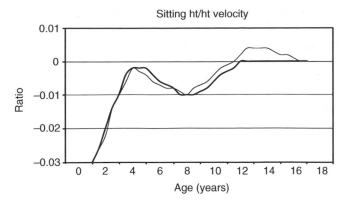

Fig. 6.E.4. Sitting height/height ratio. Mean sitting height/height ratio changes (the first derivative) for boys (thick line) and girls (thin line). The first derivatives were calculated from Gerver and de Bruin (1996)

whereas the sitting height as a fraction of the total height (Fig. 6.E.4) takes a big dip during the transition from childhood to juvenility as a measure of greater legs growth. It has recently been shown that longer lower limbs relative to body mass reduce the energetic cost of human walking (Steudel-Numbers, Weaver et al. 2007), and as the juveniles join the adult society and adult tasks, this more efficient energy economy is an obvious benefit.

The *Homo erectus* from Nariokotome,[4] also known as the "Turkana Boy," had a modern bone maturation close to that of a typical juvenile 8 "years" old and a dental age of a juvenile 8 or 10.5 "years" (by two independent estimates). His height and weight were estimated to be 160 cm and 48 kg (Dean 2007), respectively, giving him an amazing adult height prediction of 221 cm (using modern height prediction tables), as compared with nine adult *Homo erectus* fossils who had a mean height of 162 cm and weight of 54 kg. Unless a greater proportion of adult body mass and stature had been attained at an earlier age than would be expected for a modern human, the growth curve was more like that of the chimpanzee (Dean 2007).

F. TRADE-OFFS FOR THE TIMING OF TRANSITION TO JUVENILITY

Girls with premature pubarche are more inclined to develop ovarian hyperandrogenism, hyperinsulinemia, and dyslipidemia later in life (Ibanez, Dimartino-Nardi et al. 2000), the so-called polycystic ovary syndrome (PCOS). Ovarian hyperandrogenism is characterized in such women by clinical signs of androgen excess and by an exaggerated ovarian response of 17-hydroxyprogesterone to gonadotropin-relasing hormone (GnRH) agonist stimulation. High serum insulin and dyslipidemia in such subjects may be detectable as early as during juvenility, and worsen during pubertal development. These are commonly accompanied by unovulation from late adolescence onwards, with low serum levels of insulin-like growth factor binding protein-1 and sex hormone-binding globulin.

[4] A 1.5-million-year-old fossil KNM-WT 15000, a nearly complete skeleton that was discovered in 1984 at Nariokotome near Lake Turkana in Kenya.

These may constitute a genetic syndrome, but no unambiguous gene mutations have been found so far, and in line with the adaptive paradigm used here, it is suggested that they may well derive from developmental programming or an adaptive response within our adaptive phenotypic plasticity. This is supported by the fact that premature pubarche is mostly a girls' phenomenon and uncommon in boys. Should this be the case, the adaptive trade-off package includes high levels of circulating androgens in a woman, obesity, and insulin resistance. The latter two have previously been mentioned as components of the thrifty phenotype,[5] known from the outcome of intrauterine growth retardation. Indeed, children who are small for their gestational age have also an earlier transition to juvenility, characterized by early adrenarche, early pubarche, and early adiposity rebound.

What, then, are the environmental cues for early onset of juvenility? One of them is prenatal energy balance, which signals for a trade-off in the form of early onset of juvenility, which in turn provides a cue for the thrifty phenotype and hyperandrogenism in the female. Whereas hyperandrogenism compromises fertility, and may also act as a fitness-compromising element, masculinization of girls and women may be a valuable trade-off for the individual, her direct family, and the social group, under threatening environmental constraints (Hochberg and Etzioni 1995). Early adolescence will be a secondary trade-off for this package.

In postnatal life, the natural history of PCOS can be further modified by factors affecting insulin secretion and/or action, and most importantly, nutrition. This phenomenon may be regarded as one mechanism by which nutritional cues influence reproductive development. In girls with polycystic ovaries, the physiological hyperinsulinemia of puberty may affect the genesis of both ovarian hyperandrogenemia and anovulation. Higher than normal insulin levels, whether due to a genetic predisposition or excessive weight gain (or both), would exaggerate these potentially adverse effects. Pharmacological sensitization to insulin of low birth weight (LBW)-precocious pubarche girls reduced total and visceral fat and delayed menarche without attenuating linear growth (Ibanez, Lopez-Bermejo et al. 2008).

It was recently proposed that PCOS has its origin in fetal life (Franks, McCarthy et al. 2006). This hypothesis is based on data from animal models (rhesus monkey or sheep that have been exposed prenatally to high doses of androgen) and is supported by some clinical studies. It was suggested that, in human females, exposure to excess androgen, at any stage from fetal development of the ovary to the onset of puberty, leads to many of the characteristic features of PCOS, including abnormalities of luteinizing hormone secretion and insulin resistance (Franks, McCarthy et al. 2006).

The association of precocious juvenility, metabolic syndrome, and obesity, with or without PCOS may constitute a genetic syndrome; however, unambiguous gene mutations currently remain undetected, and I argue that it may well represent developmental programming or an adaptive response within our adaptive phenotypic plasticity, which may transmit transgenerationally. Thus, the mechanism may

---

[5]The thrifty phenotype hypothesis argues that as a result of environment limited in its supply of nutrients before birth, adaptations made by the fetus are associated with coronary heart disease, stroke, diabetes, and hypertension.

relate to epigenetic changes in gene expression rather than changes of gene frequency or modification of population homozygosity. This is supported by the fact that premature juvenility is less common in boys as compared to girls, whose evolutionary fitness is under greater pressure. A study of androgen receptor genotype and X-chromosome methylation found that a smaller biallelic mean of CAG repeats was associated with increased odds of PCOS (Hickey, Legro et al. 2006; Shah, Antoine et al. 2008). The chromosome bearing the shorter CAG allele was preferentially active in PCOS women. In some women, such heightened sensitivity may also result from preferential expression of androgen receptors with shorter alleles.

Another trade-off may relate to the effects of DHEA as it rises with the onset of juvenility. DHEA and DHEAS are now viewed as multifunctional steroids with both androgenic and protective roles in many aspects of the immune system and cellular function, proving also distinct effects on brain maturation as a neurosteroid (Campbell 2006). Elderly subjects who receive DHEA replacement therapy report improvement in physical and psychological well-being (Morales, Nolan et al. 1994). Several reports indicate that administration of DHEA to immune-compromised experimental animals result in preservation of the cellular immune response and resistance to induced infections by viruses, fungi, protozoa, Herpes virus type-2 encephalitis, West Nile, and coxsackievirus (Ben-Nathan, Lustig et al. 1992; Loria and Padgett 1992). Moreover, DHEA has a preserving effect on thymus involution. An environment of endemic diseases over long periods would have great demographic effects, and even a mild immunoprotective agent might acquire significant genetic selective advantage.

The social function of the juvenile, as he enters adult society, requires the androgenic effect of DHEA and DHEAS. Indeed, DHEAS levels positively correlate with ratings of aggression and delinquency among juvenile boys (van Goozen, Matthys et al. 1998), and girls with premature juvenility show higher levels of anxiety associated with increased DHEAS levels (Dorn, Hitt et al. 1999). Among women with adrenal insufficiency, DHEA supplementation demonstrated improved self-esteem, sexuality, and overall well-being, and decreased depression and anxiety (Arlt, Callies et al. 2000; Hunt, Gurnell et al. 2000), traits that are consistent with the newly assigned social role of the juvenile.

G. PRECOCIOUS JUVENILITY

An early adiposity rebound is observed in overweight children and is associated with an increased risk of overweight later in life, suggesting an association between the body habitus and transition age from childhood to juvenility (Rolland-Cachera, Deheeger et al. 2006). The typical pattern associated with an early adiposity rebound is a marked increase in BMI during juvenility that will exacerbate during adolescence. This pattern is recorded in children of recent generations as compared to those of previous generations, owing to the trend of a steeper increase of height as compared to weight in the first years of life. But even lean girls with precocious adrenarche have higher than control levels of insulin-like growth factor-I,

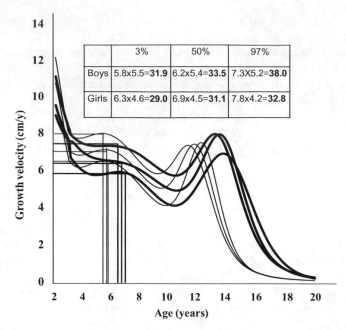

Fig. 6.G.1. The impact of age at the childhood–juvenility transition on height, comparing boys (thick lines) and girls (thin lines). The onset of juvenility is marked as the beginning of the juvenile deceleration of growth. The inset table shows the total childhood stage growth in girls and boys of the 3rd, 50th, and 97th percentile in height.

insulin-like growth factor binding protein-3 and leptin (Guven, Cinaz et al. 2005), as mechanisms that may transmit the signal of energy readiness.

As mentioned, the transition to juvenility is associated with decelerating growth (and precocious juvenility with early decelerating growth), which curbs the stable growth period of childhood. Thus, if all other life history stages remain unchanged, early juvenility compromises final adult height (Fig. 6.G.1). The 10-month earlier juvenility onset of the girl also means 10-month longer childhood of the boy. At a mean growth velocity of 7 cm/year for girls and 6.5 cm/year for boys; this delay in juvenility onset accounts for boys' height advantage of 5.9 cm. It accounts also for a difference of 6.1 cm between the 3% and 97% boys, and 4.8 cm in girls.

Several studies have searched for genetic factors that predispose children to premature juvenility in genes involved in steroid synthesis (Witchel, Smith et al. 2001; Petry, Ong et al. 2005), androgen action (Ibanez, Ong et al. 2003; Lappalainen, Utriainen et al. 2008), insulin-like growth factor-I function (Ibanez, Ong et al. 2001; Roldan, White et al. 2007), and the Wnt signaling transcription factor 7–like 2 (TCF7L2) (Lappalainen, Voutilainen et al. 2009). Although some small associations have been found, the underlying susceptibility genes remain largely unknown, or it may be a wrong assumption, with the onset of juvenility responding rather epigenetic controls of plasticity.

H. THE PYGMY PARADIGM FOR PRECOCIOUS JUVENILITY

Human pygmies are defined by some anthropologists as populations having an average male height of 155 cm. Populations exhibiting Pygmy stature reside in Africa, the Andaman Islands, Malaysia, Thailand, Indonesia, the Philippines, Papua New Guinea, Brazil, and Bolivia.[6] The small body size of human pygmies has been interpreted as an adaptation to living in dense tropical forests, thermoregulation, or endurance against starvation in low productivity environments. However, some Pygmy live outside forests in cool or dry areas; furthermore, long-standing poor nutrition does not necessarily lead to pygmy size (Migliano, Vinicius et al. 2007).

Migliano et al. constructed growth curves for the Philippine Aeta Pygmy and compared them with the lower percentiles of the U.S. growth distribution, representing undernourished individuals who grow only to average adult pygmy size (corresponding to the 0.01th percentile of the U.S. distribution). The Pygmy deviate from the U.S. undernourished sample with an early juvenile deceleration, early pubertal spurt, and early growth termination as compared to the U.S. 0.01th percentile (Migliano, Vinicius et al. 2007). In a population with a life expectancy at birth of 16 years and life expectancy at age 15 of 27 years, their first reproduction is at age 10–14 and the last reproduction averages 37.4 years. However, only 13–31% of pygmy women reached the end of their reproductive age.

According to life-history theory, age at first reproduction is set by natural selection as the result of two opposite strategies (Hochberg and Albertsson-Wikland 2008). Extended growth and large body size prompt fertility gains and reduced offspring mortality, implying a pressure for delayed reproductive onset, whereas early reproduction minimizes the likelihood of death before reproduction. Modeling fitness as a function of growth, fertility, and mortality schedules, we argued that rather than through positive selection for small stature, the short stature of Pygmy is a by-product of selection for early onset of reproduction (Hochberg, Gawlik et al. 2011). Human pygmy populations and adaptations evolved independently as the result of a life-history trade-off between the fertility benefits of larger body size against the costs of late growth cessation, under circumstances of significant young and adult mortality (Migliano, Vinicius et al. 2007).

In a recent analysis of growth during juvenility of the 1967–1969 Ju!'hoansi San data (a period when these tribes were still real hunter-gatherers) (Howell 2000), we found that their growth deceleration is by far greater as compared with modern day societies (Z. Hochberg, unpublished data). This is enhanced by the late takeoff of the adolescent growth spurt: average age 11.5 for girls and 13.5 for boys. The Ju!'hoansi San hunter-gatherers live on a protein daily intake of ~0.46 kg of meat and 2,355 Kcal. However, one of the most distinctive features of the Ju!'hoansi San people is their small body size. They are short and slender and fine-boned. Many of the people are so thin that bones and muscles are readily seen through the skin, even though most of them seem to be healthy and vigorous (Howell 2010).

---

[6]This definition is debated by those who claim that this eponym be reserved for the West and East Pygmy of Africa only.

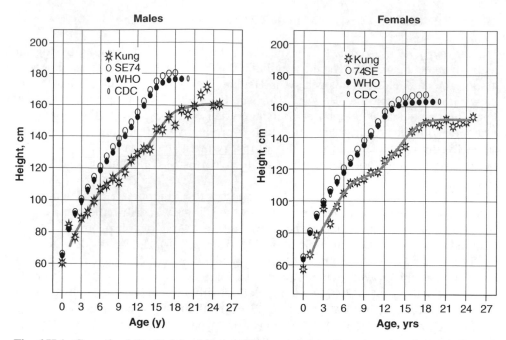

Fig. 6.H.1. Growth of the Kalahari Desert !Kung males is plotted against Swedish (SE74) WHO and USA (CDC) reference values, showing the marked deceleration during juvenility that is not recovered despite normal adolescent growth. Based on data provided by Robert Walker from the University of Missouri, Columbia, MO.

Their life history includes a mean age of marriage of 17 and 25 for females and males, respectively, first and last reproduction at 18.8 and 34.4 years, a long infancy with an interbirth interval of 41.3 months, and, as expected from their long infancy, an extremely short adult height (Howell 2000). Their probability to survival to age 1 is 0.75, to age 5, 0.7, and to age 15, 0.6 and 0.56 for girls and boys, respectively. The height SDS loss during juvenility is as vast as 1.6 for both girls and boys. Howell (2010) notes that juveniles are the thinnest members of the Ju!'hoansi San population, and the most stunted in height and wasted in body size. Their transition to juvenility, as defined by the adiposity rebound, was at 9.0 years for girls and 9.1 years for Ju!'hoansi San boys, as compared to 6.7 and 6.8 years of age for U.S. girls and boys, respectively. The juvenility deceleration is clearly visible, and its magnitude is such that despite normal adolescent growth, the height deficit is carried over to the final adult height. Preadolescent Ju!'hoansi San growth slowed for boys and girls to 2.2 and 1.9 cm/year at spurt takeoff at a mean age of 11.5 and 13.5 years, respectively, as compared to U.S. boys' takeoff velocity of 3 cm/year at age 12.8, and girls' 3.3 cm/year at the age of 10.6 (Fig. 6.H.1).

Blurton Jones and colleagues noted that Ju!'hoansi San children are much less oriented to work and obtain much less food than Hadza children do (Blurton Jones, Smith et al. 1992). They suggested that the environment of the Kalahari is more dangerous to children than that of the open hilly areas of the Hadza terrain.

Ju!'hoansi San parents worry that their children can easily become lost in the Kalahari, where it is difficult to get a long-distance perspective on the environment, and that the combination of risks from animals, snakes, and being lost is too great to justify allowing children to roam the desert seeking food.

I. EVOLUTIONARY PERSPECTIVE IN PRECOCIOUS JUVENILITY

The age at transition from childhood to juvenility is remarkably constant, especially when compared to such life-history variables as age of sexual maturation, which is subject to a wide degree of plasticity over relatively short period of time (Hochberg 2009). Comparison with African apes (but no other primates) suggests that the timing of adrenarche and the sex difference in chimpanzees out of their infancy may be similar to that in humans out of their childhood (Cutler, Glenn et al. 1978), though the full course of age-related changes in DHEAS and their relationship to reproductive and brain maturation in apes are not clear (Campbell 2006). Assuming an important role for adrenarche in human brain maturation, Campbell argued that the increased brain size and extended life span of humans relative to the great apes imply changes in the timing and impact of adrenarche (Campbell 2006). Thus, he argues that increases in body size evident among *Homo erectus* imply increases in life span and delayed reproductive maturation, and as such these are a natural point at which to start a consideration of the potential role of adrenarche in human evolution (Campbell 2006).

The syndromic precocious juvenility discussed above, and the variability in the age of adiposity rebound, imply an adaptive plasticity of no more than 2 years between early and late transition to juvenility (Cole, Freeman et al. 1995).

The transition from childhood to juvenility is closely associated with the first permanent molar teeth eruption—a landmark evident in fossil remains. A comparative study across 21 primate species found the age of first molar eruption to be highly associated with brain weight ($r = 0.98$) and a host of other life-history variables (Smith 1994).

Early adolescence, early menarche, and short stature, as reported for precocious pubarche (Ibanez, Jimenez et al. 2006), will be other trade-offs for this package. Likewise, Palmert et al. found that among 14 girls with idiopathic precocious puberty whose gonadal hormone production was suppressed by drugs, adrenarche started from age 3, and could be described from that point on as exponential (Palmert, Hayden et al. 2001). Indeed, insecure energy provision has been shown to result in short stature, as discussed above for the Pygmy, and reported for delayed infancy–childhood transition (DICT) (Hochberg and Albertsson-Wikland 2008), traded against the advantages of large body size. Early menarche is a typical trade-off of current reproduction against later under-fertility.

Early infancy and childhood growth are also strong predictors of juvenility. Allowing for current weight, children who showed rapid postnatal weight gain between the ages of 0 and 3 years had higher DHEAS and androstenedione levels at age 8 (Ong, Potau et al. 2004). Thus, juvenility onset and adrenal androgen secretion are programmed during fetal and early postnatal development, and early weight gain might therefore represent an additional mechanism that contributes to the association between LBW and hyperandrogenism.

Another trade-off may relate to the neurological effects of DHEA mentioned above, as it rises with the onset of juvenility. The social function of the juvenile, as he gains new assignments, requires the androgenic effect of DHEA. In fact, DHEAS levels positively correlate with ratings of aggression and delinquency among juvenile boys (van Goozen, Matthys et al. 1998), and girls with premature adrenarche show higher levels of anxiety associated with increased DHEAS levels (Dorn, Hitt et al. 1999). Among women with adrenal insufficiency, DHEA supplementation improved self-esteem, sexuality, and overall well-being, and decreased depression and anxiety (Arlt, Callies et al. 2000; Hunt, Gurnell et al. 2000), traits that are consistent with the newly assigned social role of the juvenile.

7

ADOLESCENCE

The terms "puberty" and "adolescence" are often incorrectly used interchangeably. For the endocrinologist, "puberty" refers to the activation of the neuroendocrine hypothalamic–pituitary–gonadal axis that culminates in gonadal maturation and sex steroids' bioeffects. "Adolescence" refers to the maturation of the juvenile into adult social and cognitive behaviors. The collective end point of these two processes is the reproductively mature adult, and, therefore, their mechanisms are interconnected. However, gonadal maturation and behavioral maturation are two distinct brain-driven processes with possible and common separate timing and neurobiological mechanisms. To ensure reproductive and parenting success in the service of reproductive fitness, hormonal and mental maturations are intimately coupled through iterative interactions between the nervous system and gonadal steroid hormones (Sisk and Foster 2004). Some, but not all, brain centers reach their final maturation during adolescence, to include the splenium of the corpus callosum for transfer of visual information, fornix for learning and memory, posterior limbic system for some sexual functions, the superior longitudinal fasciculus (SLF) for spatial attention, inferior fronto-occipital faciculus (ILF) and inferior longitudinal fasciculus (IFO) for memory functions, centrum semiovale, the genu and subcortical white matter of the gyri for some cognitive functions (Lebel, Walker et al. 2008). In the context of this discussion, the package that I call adolescence includes the growth spurt, neuroendocrine changes of puberty, the secondary sexual characteristics, and the acquisition of biosocial skills that are needed for successful reproduction.

In preparation for the complexity of human adulthood, adolescence is a period of attainment of social, economic and sexual intricacy, fecundity and fertility; this uniquely human life-history stage is characterized by pubertal development,

Evo-Devo of Child Growth: Treatise on Child Growth and Human Evolution, First Edition.
Ze'ev Hochberg.
© 2012 Wiley-Blackwell. Published 2012 by John Wiley & Sons, Inc.

including growth of the gonads and the adolescent growth spurt. In fact, the humanly unique growth spurt of adolescence has been used by some physical anthropologists and ape biologists as almost the *sine qua non* of the definition of adolescence (Bogin 1999b), even though it starts before pubertal secondary sexual characteristics in girls, and much later than the onset of genital changes in boys. At the same time, preexisting friendships build up, and new relationships develop in peer groups that facilitate intimacy and mutual support. Height and muscularity are still of youth size, preventing competition with adults, and infertility during adolescence has important implications in the social group. The unique style of human social and cultural learning requires that adolescents learn, become skilled, and practice adult-type cultural, economic, social, and sexual behaviors before reproducing.

During this "formal operational period," as defined by Piaget (Table 2.1), thoughts become more abstract, incorporating the principles of formal logic, and thinking becomes less tied to concrete reality (Piaget 1952). The ability to generate abstract propositions, multiple hypotheses, and their possible outcomes becomes evident. Formal logical systems are acquired, and the adolescent can handle proportions, algebraic operations, and other purely abstract processes. Prepositional logic, "as-if" and "if-then" steps can now be used, as well as such aids as axioms to transcend human limits on comprehension.

Boys and girls embark on different adolescent strategies to achieve their fertility goals. As menarche occurs about a year after the peak height velocity, girls have an apparent womanly body form, but they are not fertile; they will develop an adult cycle of ovulation and adult size of the birth canal at about 18 years of age. As they learn their adult social roles and while they are still infertile, they are perceived by adults as mature. The perception of fertility in girls allows them to enter the social–economic–sexual world of adult women, and to practice many skills without the risk of pregnancy (Mead 2001).

Boys will become fertile about 2 years after peak height velocity. Such adolescent boys are still young in outward development, body size, voice, and facial features. They will learn their adult social–economic–sexual roles while they are already sexually mature but not yet perceived as such by adults. This allows them to interact and learn from older adolescents and adults without seeming to compete for important resources, including women (Locke and Bogin 2006). Testosterone appears to be important for activation of the courtship behavior that leads to the formation of sexual pairing bonds (while pairing bonds in nonsexual contexts are not regulated by testosterone) (Wingfield, Jacobs et al. 1997).

There are trade-offs to rising circulating concentrations of testosterone that may mediate life history due to its pleiotropic actions. These can be energetic costs or may involve increased predation risk or reduced survival after wounding. Regulation of testosterone secretion during adolescence must balance the need to compete with other males while allowing an adjustment to the social group. In yet another trade-off, high circulating levels of testosterone for prolonged periods are also known to suppress the immune system. This latter effect may have profound implications for the development of androgen-dependent secondary sex characteristics that have evolved through sexual selection. Conservation in the actions of testosterone in vertebrates has prompted the "evolutionary constraint hypothesis," which assumes that testosterone signaling mechanisms and male traits evolved as a unit (Hau 2007). This hypothesis implies that the actions of testosterone are similar across sexes and species, and only the levels of circulating testosterone concentrations change during evolution. In contrast, the "evolutionary potential hypothesis" proposes that testosterone signaling mechanisms and male traits evolve independently. In the latter scenario, the linkage between hormone and traits itself can be shaped by selection, leading to variation in trade-off functions (Hau 2007).

The social context of an adolescent girl, such as the family composition, has important influence on pubertal timing (Matchock and Susman 2006). Absence of a biological father, the presence of half- and step-brothers, and living in an urban environment were shown to be associated with earlier menarche. The presence of sisters in the household while growing up, especially older sisters, was associated with delayed menarche. Menarcheal age was not affected by the number of brothers in the household, nor was there an effect of birth order. Body weight and race are also associated with the age of menarche. Putative human pheromones were suggested for many years to modulate sexual maturation to promote gene survival and prevent inbreeding, as occurs in rodents and nonhuman primates. Moreover, serum testosterone and estradiol correlate with sex-typical pheromone production during sexual maturation.

A. HUMAN EVOLUTION OF ADOLESCENCE

Chimpanzees progress through juvenility at 5–9 years of age, remarkably similar to modern humans, whose juvenility is followed by adolescence, and they reproduce for the first time at age 10.5 in captivity and 12.4 years in the wild. Locke and Bogin (2006) suggested that the single most important feature defining human adolescence is its skeletal component—the growth spurt, which is experienced by virtually all boys and girls. There is no evidence for a human-like adolescent growth spurt in any living ape, but with obvious limitations to interpretations obtained from skeletal remains, there is some tentative evidence that 1.8 million years ago hominids may have had a pattern of growth that suggests adolescent stage of development (Tardieu 1998).

Adolescence is also the time of subcutaneous fat deposition, mostly in girls. Whereas the subcutaneous fat of the chimpanzee female is evenly spread over her body, the human adolescent girl has striking fat deposits under the skin, especially in the area of the thighs, buttocks, and breasts, even if this individual is thin overall. Howell suggested a "fatness hypothesis" to explain this difference between the species, which asserts that prominent fat deposits evolved in humans for several reasons: (1) to provide storage of a surplus to get through periods of scarcity, as is often suggested; (2) to attract a mate as frequently noted in her research subjects, the Ju!'hoansi San of the Kalahari Desert; (3) it signals sexual condition to potential partners, so they play a role in pair-bonding and reproduction; and (4) to permit continuous monitoring of her nutritional state by others, so that food sharing can be tailored to need (Howell 2010). Loss of fur and prominent subcutaneous fat deposits allow humans to assess the nutritional state of others in a glance. Subcutaneous fat deposits serve as signals, turning on and off the impulse to provision children.

The estimated age for menarche at the beginning of the agrarian period at the end of the Pleistocene, around 20,000 to 12,000 years ago, was 7–13 years. It was estimated that full reproductive competence in New Stone Age females at 12,000 to 5,000 years ago occurred at ages 9–14. This would place menarche at 7–12 years, assuming a 2–4 year gap between menarche and reproductive competence, as recent data suggest for natural fertility nonindustrious societies (Hochberg, Gawlik et al. 2011) (Fig. 2.C.2). This suggests that menarche in Neolithic times could have been in a range even earlier as compared with menarche age observed in industrial countries. Gluckman and Hanson (2006) argued that human females evolved to enter puberty at a relatively young age to progress into reproductive competence at 11–13 years of age. Indeed, it was shown that the Aeta of the Philippines reproduce as early as ages 10–14 (Migliano, Vinicius et al. 2007)—younger than any natural fertility society that are shown on Fig. 2.A.1. This would have matched the degree of psychosocial maturation necessary to function as an adult in Paleolithic society based on small groups of hunter-gatherers, but certainly it is by far too early for the sophistication required to function as an adult in an industrial society.

Agrarian development has significantly altered the human environment, requiring adaptive evolution. Since the development of agriculture, the human population volume has grown in an exponential fashion, increasing the likelihood of mutations occurring in any gene. Agriculture also induced a more sedentary lifestyle, as people needed to live near their fields. These new cultural systems resulted in large increases in local human densities that were ideal for the spread of infectious diseases. With agriculture and life in large villages, childhood diseases and periodic famines became common, and therefore the average age of menarche was deferred as much as it is in present day for underprivileged adolescents in developing countries. It was shown that such underprivileged girls mature 2–4 years later as compared to well-off children in India, Cameroon, South Africa, and Venezuela (Parent, Teilmann et al. 2003). This matched the increasing complexity of being an adult in a society engaged in agriculture, settlement, and population aggregation, which in turn led to the differentiation of social tasks and the creation of societal hierarchies.

With habitation of small towns and then larger cities, and as hygiene deteriorated with increasing population density, the age of menarche was delayed, particularly by the turn of first millennium in Europe (Ellison 1981; Pasquet, Biyong et al. 1999). Once again, this delay matched the increased complexity of Roman, then medieval society, and even more so at the industrial revolution of the 18th century.

With modern hygiene, nutrition, and medicine, these nutritional and infectious constraints on puberty have been removed, and the age of menarche has fallen to its evolutionarily determined range. But now the complexity of society has increased enormously and psychosocial maturation takes longer. Recent studies have shown that much of our brain does not fully mature until the middle of our third decade (Lebel, Walker et al. 2008). For the first time in human evolutionary history, biological puberty—the neuroendocrine component of adolescence—significantly precedes the age of successful functioning as an adult—the mental and social component of adolescence. Gluckman and Hanson suggested that this mismatch between the age of biological and psychosocial maturation constitutes a fundamental issue for modern society (Gluckman and Hanson 2006): "Our social structures have been developed in the expectation of longer childhood, prolonged education and training, and later reproductive competence. This emerging mismatch creates fundamental pressures on contemporary adolescents and on how they live in society."

B. TRANSITION FROM JUVENILITY TO ADOLESCENCE

Age and size at adolescence have strong effects on an individual's fitness, because they affect the reproductive potential, schedule, and efficiency (Stearns 2000). Early maturing toward and during adolescence increases survival until reproduction and reduces generation time, and potentially lengthens reproductive life span. Maturing late lengthens juvenile growth, delays fecundity age, and improves offspring survival through large body-size effects. Thus, individuals face a trade-off between maturing to young reproduction or maturing to large body size, since for any given growth rate earlier maturation implies smaller size.

The onset of puberty has been the topic of much mechanistic and epidemiological research, and several excellent review articles summarized current concepts (Grumbach and Styne 2003; Parent, Teilmann et al. 2003). This will be mentioned here only briefly, as I focus on life-history evolutionary perspective and the context of the transition from juvenility to adolescence as a period of adaptive plasticity. Using the paradigm of this treatise, it speaks of a transition from juvenility to adolescence rather that an onset of puberty. This semantic matter entails two aspects that are introduced here for the first time and shed more light on what has been written up until now. Whereas "onset" implies an event that has happened over a short period of time, the claim made here is that the transition is rather a course of actions that starts during juvenility, progressively gaining power as the child enters a full-scale adolescence. The second nonsemantic matter is that during the child's transition into adolescence, hypothalamic–pituitary–gonadal maturation with its biological effects is but one aspect of the package known as adolescence. Adaptation of adolescence to the environment, the culture, and the society, and the consequences of early or late transition, involves the many facets of adolescence.

Boys mature during adolescence in a different way from girls (Table 7.1) Even during childhood and juvenility, girls have higher estrogen levels than boys (Klein, Baron et al. 1994). The onset of puberty in girls is mostly considered to take place when breast buds erupt. However, it is now recognized that this is not the first sign of maturation of the female hypothalamic–pituitary–gonadal axis. Very much like boys, who exhibit their gonads for direct palpation and show testicular growth before sex-steroid concentrations increase, the ovaries start to grow discretely about

TABLE 7.1. Adolescence in Boys and Girls Manifest Differently With Regard to Their Actual Fertility

Boys	Girls
Become fertile ~2 years after onset of growth spurt; 1 year before peak height velocity (PHV)	Complete half their breast and pubic hair development by peak height velocity; menarche ~1 year after PHV
Remain juvenile in body hair, stature, muscularity, and voice	Appear feminine, while remaining infertile
Muscle spurt and adult stature still 4 years away; ~18 years	Adult frequency of ovulation and adult size of birth canal ~ age 18 years
Learn adult social roles while fertile, but not perceived mature by adults	Learn adult social roles while infertile, but perceived by adults as mature

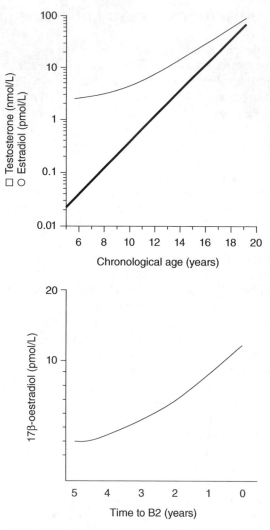

Fig. 7.B.1. Transition from juvenility into adolescence. The gradual transition from juvenility into adolescence is associated with gradual maturation of the hypothalamic–pituitary–gonadal axis, but becomes evident only when ultrasensitive assays are used. In boys, testosterone levels (thick line) increase as early as at age 6 years and estradiol levels (thin line) increase 3 years later (upper panel). In girls, estradiol levels gradually rise as of 2 years before menarche (lower panel). B2 = Tanner stage 2 for breast development Data from Ankarberg Lindgren (2005).

2 years before breast buds appear. The same is true for estradiol levels, which show a rise about 2 years before thelarche (Fig. 7.B.1) (Ankarberg Lindgren 2005), and growth accelerates at least 6 months before breast buds are evident. Boys show a similar pattern of gradual maturation of the hypothalamic–pituitary–gonadal axis. Testosterone levels start to gradually rise at a mean age of 9, long before testicular growth—the so called gonadarche (Fig. 7.B.1) (Ankarberg Lindgren 2005).

Just as much as the secular trend in human size has been an adaptive phenomenon for an encouraging environment, the receding age of adolescence and pubertal development has been an adaptive response to positive environmental cues in terms of energy balance (Fig. 2.C.2). The ever-younger age of girls' thelarche and menarche may have more than a single cause. In the past decade, the popular explanation has been that this phenomenon results from environmental exposure to endocrine disruptors that accelerate hypothalamic maturation. While it may have a bearing on the earlier age of thelarche, which is a recent trend, it can hardly explain the secular trend in the age of menarche over the past 170 years (Fig. 2.C.2).

An evolutionary perspective on this worldwide trend has been proposed by Gluckman and Hanson (2006). They have challenged the concept that this has been a disease process, and suggested that reproductive and life-history strategies might be reflected in the more frequent presentation of females with early-onset adolescence. The age at transition from juvenility to adolescence in humans has a range of physical and social correlates. Some studies show that weight and obesity are good predictors of menarche (Kaplowitz, Slora et al. 2001). Others show that height, or skeletal maturation, is more important than weight (Qamra, Mehta et al. 1991). An array of auxological parameters shows distinct changes during the transition. Leg length growth accelerates slightly in boys and decelerates in girls during the transition to adolescence (Fig. 6.E.1), whereas the sitting height growth, as a measure of spinal growth, maintains a constant growth velocity in girls and accelerates in boys (Fig. 6.E.2). When the sitting height is taken as a fraction of the total height (Fig. 6.E.3), its progression is accelerated during transition in both boys and girls, and to a similar extent.

With respect to puberty, women face a trade-off between spending a long time accumulating resources through childhood growth, thereby improving the odds for successful pregnancy, against beginning early reproduction and increasing the number of reproductive cycles. A later first birth allows for a longer period of adolescent weight gain, and heavier women in traditional societies are more fertile, both correlating with higher birth rates. This trade-off has been used to model the optimal age at first birth, which under such conditions is 18 years, near the observed mean of 17.5 years in such societies (Simondon and Simondon 1998).

In developing countries, inequality related to socioeconomic status accounts for important variations in transition age to puberty within and among countries (Fig. 7.B.2) (Pasquet, Biyong et al. 1999). In a study of 406 adolescent girls from a rural area in Senegal, the nutritional condition was estimated during infancy, childhood, and adolescence, and body composition was estimated only during adolescence (Garnier, Simondon et al. 2005). These adolescent girls were shorter and thinner than girls of the same age from developed countries, and they were less emotionally mature than girls of the same age from other developing countries. Their puberty was extremely delayed compared to that of girls of the same age from industrialized countries. The median age at the onset of breast development was 12.6 years, as compared to 9.8 years in affluent Europe, and the median age at menarche was 15.9 years (12.9 years in Europe).

The environmental cues for the transition into adolescence in humans and other mammals vary with species and gender, and it may be related to altitude, temperature, humidity, and lighting, but mostly it relates to energy balance. Sensors in the hypothalamus and hindbrain monitor these signals and permit high-frequency GnRH release when the signals reach appropriate levels (Schneider 2004). The consequences of puberty and adulthood that follow, such as the defense of territory,

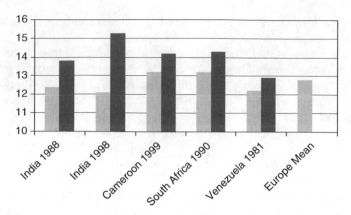

Fig. 7.B.2. Menarche age and environmental cues. Average (mean or median) menarche age in underprivileged (gray) and well-off girls (black) in different developing countries as compared to Western Europe. Data from Berkey et al. (2000). Even within countries, the age of menarche changes according to the standard of living. However, genetic difference between well-off and underprivileged populations cannot be excluded.

mating, pregnancy, and care of the young are energetically expensive. The individual must sense whether she has grown sufficiently (through metabolic cues), what her relationship is to other individuals (through social cues), and whether conditions are optimal to begin the reproductive process (through environmental cues) (Sisk and Foster 2004). For example, metabolic fuel availability and circulating concentrations of insulin, glucose, and leptin in females serve as important signals for the attainment of somatic growth that will be sufficient to support pregnancy (Warren 1983).

In the early 1970s, the direct relationship between body weight and the age at onset of puberty was suggested, and it was concluded that a critical amount of body fat was needed for the onset of puberty (Frisch, Revelle et al. 1973). Although this has been a matter of much controversy, it is in line with an evolutionary perspective of energetic cues in signaling for maturation and early menarche as an important element in enhancing reproductive fitness.

It is not clear at what age the transition from juvenility into adolescence is determined. It has been shown that the protein source of food at ages 3–5 (childhood) influences the transition age; a high animal vs. vegetable protein ratio during childhood was associated with early menarche, after controlling for body size (Berkey, Gardner et al. 2000). Consumption of more nutritious foods, such as those derived from animal protein, increased approximately 2.6 million years ago, when early hominids displayed an important behavioral shift relative to ancestral forms: the recognition that a carcass represented a new and valuable food source. The shift in the hominid "prey image" to the carcass and the use of tools for butchery increased the amount of protein and calories available, irrespective of the local scene. Life-history theory claims that 2.6 million years ago was before hominids had an adolescent stage, and a childhood stage had just started to evolve (Fig. 1.D.2); the age range of 3–5 years corresponded to childhood, and in the absence of adolescence, the transition to juvenility might have determined the age of fertility. More on juvenility in preparation for puberty has been discussed in Chapter 6, Section G.

Initial studies of the leptin–puberty connection have raised much interest, but it is now evident that leptin is but one of a series of permissive signals for the transition into adolescence. Adipokines are also involved in neuroendocrine regulation of gonadotropin secretion and the transition to puberty. Each permissive signal by itself cannot fully explain the transition into adolescence, as none of such signals is unique to this stage of life. Many are used early or later on to time other transitions from low- to high-frequency GnRH secretion, such as the seasonal variations in some animals, and restricted diet or high energy expenditure in others. Adolescence may be unique because it represents the first alignment of several permissive signals that must interplay in order to result in a maturation that never again occurs. This integrative concept suggests an innate developmental clock that times the unfolding of primary genetic programs and produces the internally derived signals, which in turn determine the responses to both internal and external permissive signals (Sisk and Foster 2004). According to this broad reasoning, permissive signals would not influence the ticking of the developmental clock, but their combination would determine precisely when the puberty alarm would sound.

In the past few decades, we have been facing a unique juvenility–adolescence transition where neuroendocrine puberty has advanced but psychosocial adolescence has been deferred to a later age. Physical growth and the hypothalamic–pituitary–gonadal axis start their transit into maturation at the end of the first decade of life, but as late as the second and early third decades of life, adolescents are not mature enough to assume adult roles. If we define adult roles as taking over responsibilities as citizens, the development of a firm partnership, living with the partner, establishing a family, caring for a family, starting a career, becoming integrated into a social group, and establishing an independent household, then European and American youth do not become adults until late into their 20s.

The phenomenon of biological maturation preceding psychosocial maturation is so clearly happening for the first time in our evolutionary history, and this "developmental mismatch," as Gluckman and Hanson describe it, has considerable societal implications. It may have a bearing on the increasing incidence of adolescent alcoholism, drug abuse, and teenage pregnancy under the influence of mature hormone and immature personality (more so in girls with precocious puberty). This is to be expected in the face of that mismatch. Sex hormones are secreted and function under ecological constraints and evolved to accommodate physiology, behavior, and their interrelationships (Wingfield, Jacobs et al. 1997). A diverse array of social behaviors is regulated by hormones, while behavioral interactions affect hormone secretion. Nonetheless, comparative field and laboratory experiments indicate that general underlying themes, including mechanisms, may exist. For example, comparative studies of birds reveal that testosterone activates a type of aggression and territorial behavior in species that are territorial only during the breeding season (Wingfield, Jacobs et al. 1997). Territoriality at other times appears to be independent of sex steroid control, although qualitatively and quantitatively the behavior appears identical. Similarly, in some populations, pairing bonds are sexual, whereas in others they appear to be alliances possibly for joint defense of a territory. In cooperative groups of birds, pairing bonds and alliances may exist simultaneously.

Testosterone appears to be important for activation of the courtship behavior that leads to the formation of sexual pairing bonds in most vertebrates. However, bonds in nonsexual contexts are not regulated by testosterone. Why this diversity in control mechanisms? It appears that there are evolutionary "costs" to high circulating levels of testosterone. They can be energetic costs or may involve increased predation risk or reduced survival after wounding. In males who express parental behavior, high circulating testosterone levels interfere with parental care, resulting in reduced reproductive success (Wingfield, Jacobs et al. 1997). Thus, regulation of testosterone secretion must balance the need to compete with other males as well as to provide parental care. High circulating levels of testosterone for prolonged periods also suppress the immune system, with possible profound implications for the development of androgen-dependent secondary sex characteristics that have evolved through sexual selection.

There are several ways to avoid potential "costs" of hormone secretion at inappropriate times. A hormone may be metabolized at its target cell (e.g., aromatization of testosterone to estradiol). Or, receptors may be downregulated in tissues that would otherwise respond inappropriately in a specific life-history state.

The juvenile–adolescence transitional slowing growth rate has been shown to be a predictive adaptive period, and the response has to do with longevity in a negative relationship with nutrition during this transition (Pembrey, Bygren et al. 2006); poor food supply results in longer survivorship. Moreover, food supply during this transition had an effect that lasted for at least the children and grandchildren generations, and interestingly enough, it was inherited from fathers but not mothers and passed to their sons and daughter (Pembrey, Bygren et al. 2006). Using the Northern Sweden Överkalix cohorts of 1890, 1905, and 1920, this study analyzed the effect of food availability on the relative risk for mortality of offspring and grandchildren, using 303 contemporary probands and their 1,818 parents and grandparents. After appropriate adjustment, the paternal grandfather's food supply was linked to the mortality relative risk of his grandsons, while the paternal grandmother's food supply was associated with her granddaughters' mortality. These transgenerational effects were observed only when exposure to poor or good nutrition took place during the juvenility–adolescence transition (both grandparents) or fetal/infant life (grandmothers) but not during either grandparent's puberty (Fig. 7.B.3). This mode of inheritance suggest that the transgenerational transmissions are mediated by the sex chromosomes X and Y.

C. PUBERTAL GROWTH

After several years of decreasing juvenile growth rate, humans have a uniquely distinct growth spurt in both sexes (Fig. 2.B.1). The onset of the spurt occurs in both sexes after a period of gonadarche, which is apparent in boys' scrotal testes and concealed in girls' intraabdominal ovaries. After a period of an "attempt" to conceal their maturation, the growth spurt is a mechanism for becoming a taller adult as rapidly as possible. However, merely continuing ordinary growth could achieve size alone; the growth spurt is a distinct pattern of achieving adult size. The growth spurt

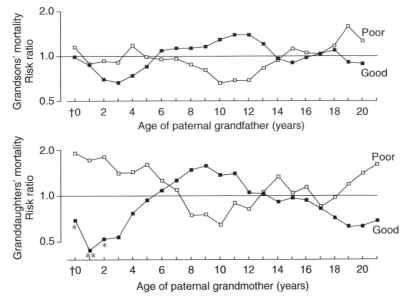

Fig. 7.B.3. Transgenerational predictive adaptive response for longevity during the juvenility–adolescence transition. The effect of paternal grandparents' food supply (good—filled squares; poor—open squares) at different times of their early life on the mortality rate of their grandchildren. During transition from juvenility to adolescence, grandfathers affect the relative risk (RR) for mortality of their grandsons (upper panel) and grandmothers affect the mortality RR of their granddaughters (lower panel). Adapted with permission from Pembrey et al. (2006).

is characteristic of all human populations, although poor nutrition can delay it or lower its dimensions. In contemporary affluent societies, the peak height velocity happens at about 13 years of age in males and 11.5 years in females.

The age of transition from juvenility to adolescence does not influence final height, other than in extreme cases of pathological precocious puberty (Bourguignon 1988). This is illustrated in Fig. 7.C.1 utilizing Karlberg's infancy–childhood–puberty model (Karlberg 1989). Despite missing the stage of juvenility, the model serves the demonstration; children who start their pubertal growth spurt at a mean age of 11 will end up as tall as those starting at age 16.

Although many other mammals gain weight at adolescence, an acceleration of linear growth is not usual, and the adolescent growth spurt as we observe it is an exclusive human trait. We have no evidence to estimate precisely when humans' adolescent growth spurt evolved. Gavan first compared relative growth curves in chimpanzees and humans: In childhood, humans fell below the chimps' curve, and then rose above it at adolescence, finally rejoining it (Gavan 1953) (Fig. 1.D.1). This work suggested that human children and juveniles might be regarded as relatively growth suppressed, with the adolescent spurt providing a compensatory catch-up growth. In terms of size, an early adolescent human male "pretends" to be more childlike than he really is. Why would such a pattern evolve? A marked spurt in height must influence entry into the adult social ladder, but the real advantage in

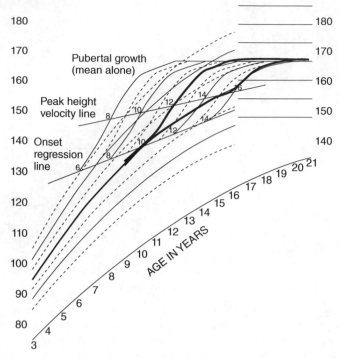

Fig. 7.C.1. The juvenility–adolescence transition according to the infancy–childhood–puberty (ICP) growth model of Karlberg (1989). The adolescent stage of the ICP model begins before the juvenile stage tapers off. The parallel lines represent the mean ±1, 2, and 3 SDS. The solid lines demonstrate how early and late transitions do not change the final height.

resembling a younger form during the preceding juvenility is thought to be the way it elicits parental behavior from adults: care, rather than competition (Weisfeld 1979). The human pattern of growth suppression followed by the spurt has the function of prolonging humans' prepubertal experience—a period of intense learning and parental care.

8

YOUTH

When adolescents in industrial societies seem to mature and gain adult body size and habitus, they mostly maintain immature behaviors and social pose. Experts in the discipline of adolescence medicine have recognized this dilemma, and expanded the age range of the subjects in their care to include young women and men until age 24. But even at 24, many of our youngsters do not assume adult roles. Indeed, many of them are too immature to assume adult tasks (Cauffman and Steinberg 2000). This study examined the influence of three psychosocial factors (responsibility, perspective, and temperance) on maturity of judgment in a sample of over 1,000 participants ranging in age from 12 to 48 years. Participants completed assessments of their psychosocial maturity in these domains and responded to a series of hypothetical decision-making dilemmas about potentially antisocial or risky behaviors. Socially responsible decision making was significantly more common among young adults than among adolescents, but did not increase appreciably after age 19 (Fig. 8.1). Individuals exhibiting higher levels of responsibility, perspective, and temperance displayed more mature decision making than those with lower scores on these psychosocial factors, regardless of age. Youths, on average, scored significantly worse than those over age 21. These findings call into question recent arguments, derived from studies of logical reasoning, that youths and adults are equally competent and that laws and social policies should treat them as such. Coming of age at 21 makes good sense.

Developmental tasks for these youths would include the following nine tasks (Seiffge-Krenke and Gelhaar 2008):

Evo-Devo of Child Growth: Treatise on Child Growth and Human Evolution, First Edition.
Ze'ev Hochberg.
© 2012 Wiley-Blackwell. Published 2012 by John Wiley & Sons, Inc.

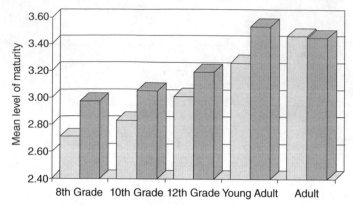

Fig. 8.1. Maturation scores of adolescents and young adults. The mean level of maturity for males (light gray) and females (dark gray) of each study group is presented. Data from Cauffman and Steinberg (2000).

1. Integration in the peer group
2. Acceptance of physical maturity
3. Establishment of an autonomous identity
4. Achievement of independence from parents
5. Preparation of future family life
6. Achievement of sociopolitical awareness
7. Preparation for an occupation
8. Having a romantic relationship
9. Formation of close friendships

On the other hand, to be considered an adult, a young woman or man may be assessed by the following eight developmental tasks for young adulthood, according to Seiffge-Krenke and Gelhaar (2008):

1. The development of a firm partnership
2. Living with the partner
3. Establishing an independent household
4. Establishing a family
5. Caring for a family
6. Starting a career
7. Becoming integrated in a social group
8. Taking over responsibility as a citizen

But even in preindustrial societies, youth is a clear life-history stage, and adolescents do not pursue fertility as soon as fecundity is established. It takes an average of 4 years before fecund adolescence reproduce (Fig. 2.A.1). The unpreparedness of adolescents for adult life was observed by Margaret Mead in her account of Samoans' coming of age (Mead 2001). Mead reports about the male youth:

> At seventeen or eighteen he is thrust into the Aumaga, the society of the young men and the older men without titles, the group that is called . . . "the strength of the village." Here he is badgered into efficiency by rivalry, percept and example. Here he has many rivals. . . . He must always pit himself against them. . . . He must become a house builder, a fisherman, an orator or wood carver . . . while reluctant to accept responsibility. Marrying a girl without proficiency would be a most imprudent step.

Likewise, the female youth do not become mature women as soon as they mature physically:

> The seventeen-year-old girl does not wish to marry—not yet. It is better to live as a girl with no responsibility, and a rich variety of emotional experience. This is the best period of her life. . . . The long expedition after fish and food and weaving materials give ample opportunities for rendezvous. Marriage is the inevitable to be deferred as long as possible. She is part of the organization of young girls and the wives of untitled men and widows, the Aualuma.

In the *Descent of Man*, Darwin tells a similar story:

> Savages almost always marry; yet there is some prudential restraint, for they do not commonly marry at the earliest possible age. The young men are often required to shew that they can support a wife; and they generally have first to earn the price with which to purchase her from her parents. With savages the difficulty of obtaining subsistence occasionally limits their number in a much more direct manner than with civilised people, for all tribes periodically suffer from severe famines. At such times savages are forced to devour much bad food, and their health can hardly fail to be injured.

At the physiological domain, the youths continue to mature in many different aspects. Their growth hormone levels will peak at 20 (Zadik, Chalew et al. 1985). Energy expenditure and physical activity level will increase until age 21 (Lantz, Bratteby et al. 2008), their bone mineral density until age 22 and beyond by some records. (Teegarden, Proulx et al. 1995). Circulating testosterone in youth will rise until age 23 (Uchida, Bribiescas et al. 2006).

Brain development continues to mature (Lebel, Walker et al. 2008), and fractional anisotropy of white matter tracts (by fiber tractography), continues to grow for the corticospinal tracts until age 20, for inferior fronto-occipital tracts until age 22, the globus palidus to age 23, cingulum and putamen until age 25, and for the uncinate fasciculus to age 30 (Fig. 8.2.)

Youth is not a recent life history stage. The chimpanzee does not reproduce until 2–3 years after completing her puberty (Leigh and Shea 1996). The female will have adult-size labial swelling at 10 and menarche at 11, but will reproduce for the first time in the wild at age 14 or 15. Age 14 in the female and 15 in the male is also the time when they will cease growing (Hamada, Udono et al. 1996). The male will ejaculate at age 9 and have an adult-size scrotum and testes at 12–13. His mating is flexible and comprises three distinct mating patterns (Tutin 1979): (1) opportunistic, noncompetitive mating, when an estrous female may be mated by all the community males; (2) possessiveness, when a male forms a special short-term relationship with an estrous female and may prevent lower-ranking males from copulating with her; and (3) consortships, when a male and a female leave the group and remain alone, actively avoiding other chimpanzees.

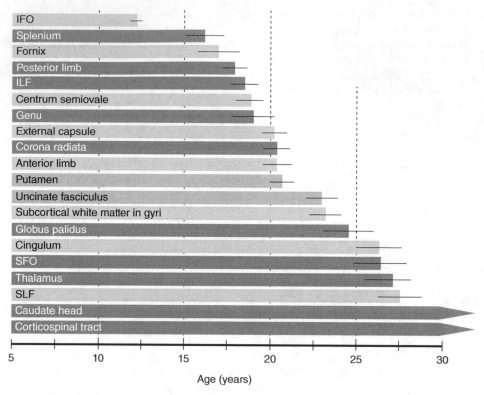

Fig. 8.2. Age at brain structures maturation. Data from Lebel et al. (2008).

It was previously noted that eruptions of the first molar teeth coincide with the childhood–juvenility transition and that of the second molar with the juvenility–adolescence transition. I suggest here that for the purpose of comparative life history and paleoanthropology, eruption of third molars be considered a marker of full maturity and transition from youth to adulthood. The chimpanzee will erupt her third molars at age 10.5 years in captivity, but only at 12–14 years in the wild. Modern humans do so at age 20–21, which may signify full maturation and transition to adulthood.

9

EVOLUTIONARY STRATEGIES FOR BODY SIZE

There are advantages and disadvantages in being big or small, and selection for size is, almost every time, for an optimum as dictated by the environment and the species-specific size strategy. Growth utilizes all four levels of human biological evolutionary adaptation. These include changes of gene frequency in a population or species, which exerts its impact over millennia, modification of population homozygocity, with impacts over centuries, plasticity that acts over the total life cycle of the individual and three to four further generations, and short-term acclimatization, such as in case of drought or temporary famine. Thus, body size and its response to environmental cues may be quite rapid.

From Darwin's *The Descent of Man* (1871):

> In regard to bodily size or strength, we do not know whether man is descended from some small species, like the chimpanzee, or from one as powerful as the gorilla; and, therefore, we cannot say whether man has become larger and stronger, or smaller and weaker, than his ancestors. We should, however, bear in mind that an animal possessing great size, strength, and ferocity, and which, like the gorilla, could defend itself from all enemies, would not perhaps have become social: and this would most effectually have checked the acquirement of the higher mental qualities, such as sympathy and the love of his fellows. Hence it might have been an immense advantage to man to have sprung from some comparatively weak creature.

With growth as the focus of this treatise, it is interesting to follow the continuous tendency of the stature of hominids for bigger size (Fig. 9.1) (Wang and Crompton 2003). Over approximately 4 million years, the increase in size has been about 45 cm

Evo-Devo of Child Growth: Treatise on Child Growth and Human Evolution, First Edition.
Ze'ev Hochberg.
© 2012 Wiley-Blackwell. Published 2012 by John Wiley & Sons, Inc.

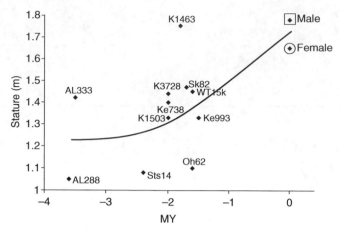

Fig. 9.1. Fossil data on tendency of hominine stature over evolutionary times. The fossil specimens are identified by the reported codes. Adapted with permission from Wang and Crompton (2003).

in stature and 30 kg in weight. This trend is in line with Cope's rule[1] asserting that population lineages tend to increase body size over geological time, and this renders the clade more susceptible to extinction. The classical example is that of the Equidae clade (horses), with small horses evolving into larger ones. Thus, the grand strategy of the hominids has been to grow bigger, in parallel to long survivorship, late maturation, and first reproduction.

The reasons for hominids to evolve a bigger size are not quite clear, but most probably they were related to the changing environment, which provided access to richer foods and favored stronger bodies for big-game hunting. About 30,000 years ago, hunter-gatherer Cro-Magnon[2] humans reached a peak male height of 174–178 cm, with men being the same 10–15 cm taller than females as they are today.

With the subsequent Ice Age and scarcer big game, many populations switched to farming and lesser quality food; great body strength was not necessarily a premium. Farmers adopted a sedentary life style, and population density increased. The impact of population density on size is well established for all species. Figure 9.2 shows the impact of numerical abundance of plankton on their size (Cohen, Jonsson et al. 2003). The same is true for humans; smaller body sizes are found in high-population-density contexts, presumably because of increased nutritional constraints and disease loads. In addition, there is evidence of mortality-based selection for relatively faster/earlier ontogeny in small-bodied hunter-gatherers living at high densities (Walker and Hamilton 2008). This was interpreted as an evolved reaction norm for earlier reproductive maturity and consequent smaller adult body size in high-mortality regimes, as discussed in Chapter 6, Section H.

[1] Edward Drinker Cope (1840–1897), American paleontologist and comparative anatomist.
[2] Humans living 10,000–40,000 years ago in the Upper Paleolithic period of the Pleistocene era. They were named after the cave of Cro-Magnon in southwest France, where the first specimen was found.

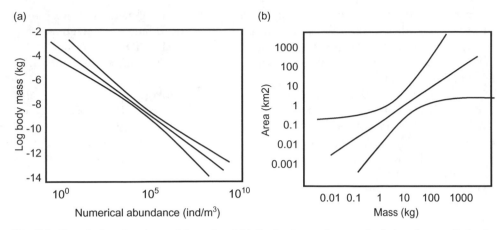

Fig. 9.2. Population density and boy size. (A) Body size and numerical abundance of plankton in Tuesday Lake, MI. Data from Cohen et al. (2003). (B) Body size dependence on individual area use in mammals. Data from Jetz et al. (2004).

But hunter-gatherers of southern Africa are all of small body size. The small bone size of Holocene foragers of the Cape Ecozone, southern Africa, are of people who are presumably the ancestors of the contemporary Ju!'hoansi San, who measure 160 and 150 cm for the average man and woman, respectively (Pfeiffer and Sealy 2006). Surely some of the environments in which these people lived must have been rich, and some of the centuries must have been times of plenty, yet the bones are always small compared to Holocene Europeans. Short stature may not be an immediate reaction to the poor quality of the environment where they lived. Howell (2010) believes that for the Ju!'hoansi San, small people may be more attractive to the opposite sex. They told her that the tallest women are considered unattractive. Shorter men tend to have somewhat better hunting success than tall men, and hunting success is attractive to women (Lee 1976; Lee and Devore 1976).

Small size at the Kalahari desert may be matter of the "cost" of food (Howell 2010). Although there is no shortage of food species in the local environment; the people work relatively short hours to acquire food, which seems to imply that food is not the limiting factor in body size (Lee 1976; Lee and Devore 1976). But perhaps food is so "expensive" for adults to go find and collect it, carry it back, process it, and consume it that people aim to consume the minimum amount consistent with survival, not the amount that produces maximal growth and large body size (Ulijaszek 2001). In another context, Howell (2010) noted that "the foods of the veldt are frequently bitter, fibrous, and difficult to process. Adults and children are unlikely to overeat such foods." Adults and even children of the Ju!'hoansi San smoke tobacco as often as they can get it, which is likely to contribute to lower appetite and to health problems, perhaps even stunted growth (Howell 2010). And everyone walks long distances through the hot desert sands and generally shows a level of physical strength and endurance that others are hard pressed to match.

Farmers of the population-dense agrarian period became shorter, and then much shorter when they urbanized and industrialized. These changes utilized genetic means—changes in gene frequencies, and we recently learned that hundreds of variants cluster in genomic loci and biological pathways and affect human height (Allen, Estrada et al. 2010).

About the year 1800, heights took a dip as industrial towns became polluted and overcrowded. Whereas this has been a process of evolutionary biology in its gradual cumulative selection by small steps of genetic adaptation to the environment through modification of population homozygocity (Roff 2007), the gradual secular trend in growth and puberty since about 1840 (Arcaleni 2006) can hardly be imagined to result from gene base sequence modifications. The worldly trend represents the impact of environmental forces applied to an inherent adaptive plasticity, which by itself is inherited through classical genes genetics. This process is the essence of evolution, as detailed in Chapter 1, Section G.

The mechanism underlying the secular trend in growth and puberty is only partly understood. Contemplating the changes after the industrial revolution into modern and postmodern life, children grow up in a totally different environment than our immediate ancestors, in families that have totally different lifestyle. A study of factors associated with the secular trend in height reported the following variables to be associated with height in most groups: maternal and paternal height, child's birth weight, maternal age at child delivery, child's length of pregnancy, and ethnic background (Rona and Chinn 1982). Modern families are smaller than in any of the natural fertility societies, and indeed children grow better in smaller families, and their size is sequentially smaller in families of four or more children (Moyes 1976). Interestingly, the association of short stature with large family size concentrates in the older members of the sibship and is independent of birth interval (Moyes 1981). Association between birth interval and stature was most pronounced for the spacing following the index child, less for the preceding interval, and least for the family average interval. It was argued that these results were compatible with deprivation of parental care and love, for which older children have to compete with their younger siblings (Moyes 1981).

Emotional deprivation sufficient to retard growth is not confined to extreme circumstances or to broad social groups, but occurs selectively within the heart of the normal family. Variables that were associated with height almost exclusively in the inner cities included parents' employment status, maternal social class, maternal employment and hours of work, and receipt of social benefits and school meals (Rona and Chinn 1982). Interestingly, the father's social class, the maternal education, or one-parent families were not associated with height.

In the animal kingdom, selection for very large sizes may be detrimental. Over the past 50 million years, successive clades of large carnivorous mammals diversified, grew bigger, and then declined to the point of extinction. Just as cars were made bigger until a fuel crisis emerged and they became smaller again, energetic constraints and pervasive selection for larger size in carnivores lead to dietary specialization (hypercarnivory) and increased vulnerability to extinction. In two major clades of extinct North American canids,[3] the evolution of large size was associated with a dietary shift to hypercarnivory and a decline in species durations to about 6

---

[3] Dogs and foxes, including wolves, coyotes, jackals, etc.

million years before becoming extinct, which are considerably shorter than for clades of smaller creatures and a more diverse diet (Van Valkenburgh, Wang et al. 2004).

The body size of animals can only be understood in the context of evolution. To evolve, growth traits must offer a selective advantage, even if small and incoherent, and this advantage will usually link to energy resources on the one hand, and to selection advantages in terms of reproduction, defense, and social competition on the other. In lower species, average body size correlates negatively with numerical abundance and positively with trophic height[4] (Brown and Gillooly 2003). A plot of body size against the available life area for mammalian species shows that humans have clearly deviated into a scale of their own (Jetz, Carbone et al. 2004) (Fig. 9.2). Using life-history theory, it is inherent that selective advantages operate at all stages of development, and not only on the adult outcome, raising the probability of reaching adulthood and reproduction.

A. THE LITTLE PEOPLE OF FLORES

It was mentioned that bigger is not necessarily better under all environmental conditions. Islands and islanders have their own unique ecology; the "island rule" implies the phenomenon of miniaturization of large animals, and the gigantism of small animals living on islands (Raia and Meiri 2006). Whereas the evolutionary events leading to smaller animals on islands are not clear, it was suggested that the extent of dwarfism on islands depends on the existence of competitors and, to a lesser extent, on the presence of predators.

In 2003, paleoanthropologists made a discovery on the Indonesian island of Flores that shook the world of early human studies like no other in recent years. Inside a cave, under a thick blanket of sediment, they unearthed 18,000-year-old fossil bones of a 1-m tall, 25-kg heavy adult human with a remarkably small head. Apparently, these hominids inhabited Flores 95,000 to as recently as 13,000 years ago, which means it would have lived at the same time as modern humans, who might have met them (Brown, Sutikna et al. 2004). The bones were unearthed during a dig at Liang Bua, a limestone cave deep in the Flores jungle. After much debate, it is now felt that the fossil appeared not to have been that of a diseased dwarf *Homo sapiens*, but rather of an entirely new species, which its discoverers from the very beginning named *Homo floresiensis*. Long arms, a sloping chin, and other primitive features suggested affinities to ancient human species such as *Homo erectus*.

Cut off from the rest of the world on this island, the species (nicknamed the "hobbit") evolved a small stature as an adaptive phenomenon, much like the pygmy elephants it is thought to have hunted. Sophisticated stone tools found nearby suggest they were not lacking in intelligence, even though the "hobbit" specimen's brain size of a mere 400 cubic cm was no larger than a chimpanzee's. This relationship suggests that the little people of Flores had been isolated on this island for 800,000 years by themselves, genetically cut off from the rest of the world, where very few other animals could get to. A small body size of such an islander is an

---

[4] Position on the food chain: what it eats and what eats it.

example of adaptive advantage under certain environmental cues. Small is efficient when one lives in dense vegetation, when thermoregulation is a challenge, or when one lives in low productivity environments.

B. LESSONS FROM THE GREAT APES

An important lesson on hominids' body size is drawn from studies of the great apes, which exhibit only two youth life-history stages: infancy and juvenility (Leigh and Shea 1996). Size variation in African apes—the Western gorilla (*Gorilla gorilla*), the bonobo (*Pan paniscus*), and the chimpanzee (*Pan troglodytes*)—is substantial, both within and between species. The evolutionary significance of this variation was analyzed through the ontogeny of size variation in this group. Intergeneric variation in size (hybrids of different genera) is largely a consequence of differences among species in the rate of body weight growth, whereas interspecific size variation (hybrids among organisms of different species are sterile) in *Pan* is a product of both rate and duration of each growth stage.

Size-related ontogenetic variation can be best understood with respect to ecological risks (Leigh and Shea 1996). Growth rates and adult size correlate negatively with ecological risk in African apes (Janson and van Schaik 1993), as they do in humans, suggesting links between ontogenetic patterns, social, and ecological variables. High growth rates in gorillas, compared to the chimps, are most consistent with this model. The same is true for the variation between common chimpanzees and pygmy chimpanzees (especially females) that seem to fit predictions of this risk model.

Utilizing life-history theory, the growth-related strategy includes the adding or deleting of life-history stages; as mentioned, humans added a childhood stage, changing the rate of growth within each stage, shortening or prolonging stages. There is no adolescent growth spurt in nonhuman apes, and a shorter infancy than any other ape is evident in humans. We have seen that all three options have been utilized during evolution.

As a model to help understand growth patterns within a species, sexual dimorphism among apes is an interesting paradigm to study. It is largely the result of sex differences in the duration of body growth in gorillas and pygmy chimpanzees, but in common chimpanzees sexual dimorphism results from differences in the rate of growth. The great degree of ontogenetic variation within and among these species in the timing and magnitude of growth during life-history stages suggests hormonal impacts.

C. THE HANDICAP THEORY

Amotz Zahavi's "handicap theory" addressed sexual selection when the costliness of the male character, such as a peacock's tail, for example, is positively attractive for the female, but may be hazardous to the male other than during mating (Zahavi 1977). When males vary in their qualities, some of the males may possess an apparent handicap: a costly feature, which reduces survival, but may increase reproduction. If only males with high-quality genes survived despite their handicap, the female must have recognized the handicap as a quality character. The handicap has no advantage in the case of nonsexual bonds. Only in reproduction does it assure the female that the male has a good genotype and is not deceitful. She has an advantage if she uses the apparent handicap for choosing a partner, and in consequence, he has an advantage if he displays it.

Using the handicap theory, under environmental conditions of energy shortage, tall stature may become a handicap. If energy-proficient short males are ignored by females and are less successful in reproductive fitness, males will be taller despite the handicap. You can see it also in the shorter life span of the males.

In fact, tall stature has a clear advantage in survival; on an evolutionary time scale, the tall live longer. Using bioarchaeology of skeletons, it was shown that both sexes display a statistically significant inverse relationship between adult height and age at death, and more so in males than in females (Kemkes-Grottenthaler 2005). Taking an epidemiological approach, the risk model implies that the estimated odds of survival beyond age 40 improve by approximately 16% for 1 SD in bone length. Overall, the relationship between body height and longevity is not causal but coincidental: mitigated by diverse environmental factors such as nutrition, socioeconomic stressors, and disease load.

But does size matters on the human mate market? Both stated preferences and mate choices have been found to be nonrandom with respect to height and weight. But how universal are these patterns? Most of the literature on human mating patterns is based on postindustrial societies. Much less is known about mating behavior in more traditional societies. In a study of a forager community—the Hadza of Tanzania—there was no evidence for assortative mating for height, weight, BMI, or percentage fat; neither was there evidence for a male-taller norm, or that number of marriages was associated with size variables (Sear and Marlowe 2009). Hadza couples may assort positively for grip strength, but grip strength did not affect the number of marriages. This lack of size-related mating patterns might appear surprising, since size is usually assumed to be an indicator of health, productivity, and overall quality. But health and productivity may be signaled in alternative ways in the Hadza, who are a small, relatively homogeneous population. There may be some disadvantages to a large size in a food-limited society where the costs of maintaining a large size during periods of food shortage may be high. Such disadvantages will not be seen in food-abundant societies, so that a large size may be a better indicator of quality in Western populations. The authors conclude wisely that "it is time to expand our horizon to a truly cross-cultural view and begin to sort between highly variable and truly universal mate patterns" (Sear and Marlowe 2009).

D. SEXUAL DIMORPHISM

Sexual selection has played an important part in differentiating the evolution of man, and an inherent part of it is sexual dimorphism—how women are different from men. Charles Darwin wrote the following in his concluding chapter of *The Descent of Man*:

> Sexual selection depends on the success of certain individuals over others of the same sex, in relation to the propagation of the species; whilst natural selection depends on

the success of both sexes, at all ages, in relation to the general conditions of life. The sexual struggle is of two kinds; in the one it is between individuals of the same sex, generally the males, in order to drive away or kill their rivals, the females remaining passive. Whilst in the other, the struggle is likewise between the individuals of the same sex, in order to excite or charm those of the opposite sex, generally the females, which no longer remain passive, but select the more agreeable partners.

The size of a male or a female organism reflects its evolutionary needs, but also a preference by the opposite sex. Darwin quotes a certain M. Carbonnier in regard to size, who maintained that the female of almost all fishes is larger than the male, and that tortoises and turtles do not offer well-marked sexual differences, whereas the males of many birds are larger than the females, and that this is the result of the advantage gained by the larger and stronger males over their rivals. With mammals and the marsupials of Australia, the sexes differ in size, and the males are almost always larger and stronger. The most extraordinary case is that of one of the seals (*Callorhinus ursinus*); a full-grown female weighs less than one-sixth of a full-grown male. From Darwin:

> The law of battle for the possession of the female appears to prevail throughout the whole great class of mammals. Most naturalists will admit that the greater size, strength, courage, and pugnacity of the male, his special weapons of offence, as well as his special means of defence, have been acquired or modified through that form of selection which I have called sexual.

> With mankind the differences between the sexes are greater than in most of the Quadrumana,[5] but not so great as in some, for instance, the mandrill. Man on an average is considerably taller, heavier, and stronger than woman, with squarer shoulders and more plainly-pronounced muscles. . . . Man is more courageous, pugnacious and energetic than woman, and has a more inventive genius. His brain is absolutely larger, but whether or not proportionately to his larger body, has not, I believe, been fully ascertained. In woman the face is rounder; the jaws and the base of the skull smaller; the outlines of the body rounder, in parts more prominent; and her pelvis is broader than in man (3. Ecker, translation, in 'Anthropological Review,' Oct. 1868, pp. 351–356. The comparison of the form of the skull in men and women has been followed out with much care by Welcker); but this latter character may perhaps be considered rather as a primary than a secondary sexual character. She comes to maturity at an earlier age than man. As with animals of all classes, so with man, the distinctive characters of the male sex are not fully developed until he is nearly mature; and if emasculated they never appear. The beard, for instance, is a secondary sexual character, and male children are beardless, though at an early age they have abundant hair on the head. It is probably due to the rather late appearance in life of the successive variations whereby man has acquired his masculine characters, that they are transmitted to the male sex alone.

> Male and female children resemble each other closely, like the young of so many other animals in which the adult sexes differ widely; they likewise resemble the mature female much more closely than the mature male. The female, however, ultimately

---

[5] An obsolete term for "apes with four hands," comprising apes, monkeys, baboons and lemurs, but not humans who are "Bimana," with two hands and two legs.

assumes certain distinctive characters, and in the formation of her skull, is said to be intermediate between the child and the man. (4. Ecker and Welcker, ibid. pp. 352, 355; Vogt, 'Lectures on Man,' Eng. translat. p. 81.) Again, as the young of closely allied though distinct species do not differ nearly so much from each other as do the adults, so it is with the children of the different races of man. Some have even maintained that race-differences cannot be detected in the infantile skull. (5. Schaaffhausen, 'Anthropolog. Review,' ibid. p. 429.) I have specified the foregoing differences between the male and female sex in mankind, because they are curiously like those of the Quadrumana.

Sexual size dimorphism may have evolved in order to increase the time available for growth in the larger male. The gorilla requires a 3/2 ratio of male/female size, whereas in humans, the male is but 8% taller and 16% heavier than the female and takes longer to mature. He will therefore mature to fecundity and reproduce later than the female. As the brain grows, sexual size dimorphism may not be as essential, and it is the human female brain, which has bridged the gender gap by growing more than the male did in hominine evolution.

It is also the female who needs to exploit any selection advantage to extend her fecundity period through early maturation, which is associated by shorter stature. Whereas height does not greatly influence fertility, it has a significant effect on offspring mortality (Allal, Sear et al. 2004). Thus, age at first birth represents a trade-off between time allocated to growth and greater fertility when mature. The heterochrony of growth and maturation is hereby realized: a uniform stretching or translation of the growth trajectory along the time axis that closely superimposes on the maturation trajectory.

Again from Darwin's perspective in *The Descent of Man*:

With mankind, especially with savages, many causes interfere with the action of sexual selection as far as the bodily frame is concerned. Civilised men are largely attracted by the mental charms of women, by their wealth, and especially by their social position; for men rarely marry into a much lower rank. The men who succeed in obtaining the more beautiful women will not have a better chance of leaving a long line of descendants than other men with plainer wives, save the few who bequeath their fortunes according to primogeniture. With respect to the opposite form of selection, namely, of the more attractive men by the women, although in civilised nations women have free or almost free choice, which is not the case with barbarous races, yet their choice is largely influenced by the social position and wealth of the men; and the success of the latter in life depends much on their intellectual powers and energy, or on the fruits of these same powers.

In the case of the chimpanzee, its life history is on the brink of extinction. With a menarche at age 10 years, first delivery at 12, a long infancy of 5 years (and hence a cycle of 5–6 years between pregnancies), and survival in the wild to 25 that translates to 13 adult/fertility years (Hill, Boesch et al. 2001), she can hardly squeeze two to three offspring into her life history. It was discussed earlier that hominids overcame this reproductive limit by a shorter infancy and inserting childhood between infancy and juvenility. Selection has resulted in greater differentiation among females, who strive to accomplish as many fertility cycles as possible, than among males. Consequently, variation among females in the ontogeny of body size may be associated with important differences in female life-history strategies.

In 2009, a special issue of *Science* was devoted to describe the eldest hominid *Ardipithecus ramidus*, who roamed what is now Ethiopia 4.4 million years ago (White, Asfaw et al. 2009). The most complete skeleton of a female was nicknamed "Ardi," who lived more than a million years before "Lucy." Ardi, who weighed about 50 kg and stood about 120 cm tall, had a mix of "primitive" traits, shared with her predecessors, the primates of the Miocene epoch, and "derived" traits, which she shared exclusively with later hominids. She was a mosaic creature, that was, neither chimpanzee nor human, living in a woodland environment, where she climbed on all fours along tree branches but walked, upright, on two legs, while on the ground. Interestingly enough, *Ardipithecus ramidus* had minimal skull and body size dimorphism, most likely associated with relatively weak male–male agonism in a male philopatric social system.

Males and females might be expected to show similar growth trajectories until juvenility, since their functional requirements are apparently similar prior to sexual maturation. Indeed, early-growing parts of the skeleton, such as facial bones, are less sexually dimorphic than long bones that grow later (Humphrey 1998). Yet the growth charts reveal a significant degree of sexual dimorphism as early as the first months of life (Fig. 4.C.1), and substantial dimorphism during juvenility and adolescence. It is quite obvious that after a short period of about 6 months, when male infants grow significantly faster than female infants, the remainder of infantile growth has minimal sexual dimorphism, if any. The transition from infancy into the childhood growth stage is also sexual dimorphic, and happens slightly earlier, by about 1 month in the female. Derived from the second derivative growth charts, childhood is shorter and the transition into juvenility happens by about 6 to 10 months earlier in the female, corresponding to the gender-specific social assignment of the juvenile in traditional societies, where boys and girls assume different functions. If the transition to juvenility is to be defined by the transition from an accelerating into a decelerating pattern in the second derivative growth curve, it suggests a mean onset age of 4.5 years in girls and 5.5 years in boys.

Interestingly, the transition age to juvenility by this definition is quite constant in girls, whereas boys show a range between ages 5 and 6. If the transition to juvenility starts with adiposity rebound, the mean age would be 5.2 years in girls and 6 years in boys. During juvenility, both genders decelerate their growth rate, yet girls start this stage by a quite marked "mid-childhood" spurt, followed by a rapid deceleration, whereas boys show a continuous milder deceleration.

If the levels of adrenal androgens are to be considered the definition, then juvenility was suggested to set in girls at ages 6–7, and in boys at ages 7–8 (Sizonenko 1978). However, recent data show an onset for boys at age 6.5, when girls of that age already show higher DHEAS levels (Cano, de Oya et al. 2006).

Sexual dimorphism is obvious in the onset and progress of adolescent growth. Except in the extreme, energy imbalance affects reproductive maturation and fertility relatively less in males, perhaps because males do not face the cost of pregnancy and lactation (Schneider 2004).

Calculating first and second derivatives and sexual dimorphism highlight gender differences in many of the auxological variables, as shown here for subischial leg length, sitting height and bi-iliac diameter (see Chapter 6, Figs. 6.E.1–6.E.4).

E. THE ROLE OF SEX STEROIDS

Sex steroids mediate male vs. female trade-offs in many different ways (Hau 2007). Testosterone increases male reproductive success by promoting courtship and sexual behaviors, territorial aggression, secondary sexual characteristics and sperm production, while often simultaneously decreasing fitness by suppressing traits such as immune function and parental care. Such pleiotropic antagonistic actions of testosterone on male traits might be an important part of the mechanistic cascade that mediates reproductive trade-offs. In evolutionary terms, the linkage between testosterone and male traits are inseparable, while both maintain inherent evolutionary adaptive plasticity.

Testosterone per se is only one of several options available for androgenic evolutionary trade-offs and their adaptive plasticity. Conversion by 5-alpha reductase to a super agonist form of the androgen receptor results in the generation of dihydrotestosterone, with a 20-fold greater receptor affinity. Aromatase-dependent conversion allows testosterone and other weaker androgens to generate estrogens, while losing in androgenic action. The biological actions of both androgenic and estrogenic steroids are further modified by steroid-binding proteins, affecting bioavailability of the hormones at target cells, steroid receptor transactivators and modulators, altering the genomic actions of the hormone-receptor complex and other downstream mechanisms modulating the tissue response. Thus, phenotypic plasticity has a wide range of variables to modify, creating a multitude of optional phenotypes.

Selection pressures favoring a control of aggressive behavior by testosterone in males could have led, via correlated evolution, to a corresponding linkage between testosterone and aggressive behavior in females. Indeed, in many vertebrate species, females are as aggressive as males with comparable peak testosterone levels in males and females during the breeding season. Furthermore, females possess androgen receptors in brain areas associated with aggressive behavior, and experimental administration or disease-related overproduction of testosterone often enhances female aggressive behavior.

It was suggested that the regulation of male traits by testosterone has important implications for the evolution of vertebrate life histories (Hau 2007). Different seasonal activities, such as reproduction or molting (shedding of old feathers in birds or old hairs in mammals), necessitate the seasonal adjustment of the individual's allocation of resources. An important mechanism of resource redistribution is via variation in plasma testosterone concentrations. Indeed, the majority of male vertebrates show large seasonal variations in plasma testosterone concentrations, with elevated levels during the reproductive season and low (or nondetectable) levels during the nonbreeding season. Important experiments revealed that vertebrates could seasonally alter the linkage between testosterone and certain life-history traits. Western song sparrows vary the endocrine control of aggressive behavior between different seasons (Soma, Sullivan et al. 1999). During the breeding season, testosterone increases the frequency and intensity of male aggressive behavior. However, during the nonbreeding season, when the gonads recede and plasma testosterone concentrations are low, territorial aggression is regulated by estradiol. Estradiol at this time of the year is unlikely to come from the receding testes, but instead it is generated within the brain from adrenal DHEA. DHEA concentrations are elevated in song sparrows during the nonbreeding season, and administration of DHEA stimulates song during aggressive encounters.

In humans, both the adrenal gland and the ovaries generate androgens, with the former producing DHEA and DHEAS, providing a tonic low-grade androgenic activity, and the latter producing the potent androgen testosterone that is secreted in relation to the menstrual and reproductive cycles.

10

ENERGY CONSIDERATIONS

Evolutionists have become increasingly interested in framing the study of hominid evolution in an ecological perspective (Leonard and Robertson 1997). Major trends in human evolution were precipitated by large-scale ecosystem changes, and research on bioenergetics has become a central component of ecosystem ecology (the biological transfer and utilization of energy). These disciplines now examine how individuals and populations extract energy from their environment, and how, in turn, that energy is allocated for biological processes such as maintenance, activity, growth, and reproduction.

The connection between size and energy availability have long been appreciated. Young (i.e., small) organisms respire more per unit of weight than old (large) ones of the same species because of the overhead costs of growth, but small adults of one species respire more per unit of weight than large adults of another species because a larger fraction of their body mass consists of structure rather than reserve; structural mass involves maintenance costs, reserve mass does not. Hence, as a general rule, smaller species have lower total requirements consistent with their smaller size. Yet, their rate of energy use per unit body mass is increased. Kleiber (1932) showed that species differences in basal metabolic rate (BMR) are linked to body size, with total BMR scaling as a power function of body mass with an exponent of three-quarters, and with mass-specific BMR scaling with an exponent of –0.25 (Fig. 10.1). Such "quarter-power" scaling, known as Kleiber's Law, seems to arise from the fractal geometry of the vasculature that supplies cells with metabolic substrate (West, Brown et al. 1997). The role of thermogenesis in metabolism remains unclear, in part because Kleiber's Law, as originally formulated, was based upon the idea that metabolic energy was entirely related to measurements of heat generation and loss.

Evo-Devo of Child Growth: Treatise on Child Growth and Human Evolution, First Edition.
Ze'ev Hochberg.
© 2012 Wiley-Blackwell. Published 2012 by John Wiley & Sons, Inc.

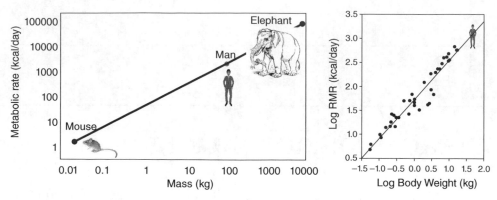

Fig. 10.1. Kleiber's Law: Left panel: The log basal metabolic rate of mammals varies with log body mass as a straight line with a slope of 0.74. (3/4 scaling relationship; 1 kg → 0.74 Kcal). Right panel: Log-log plot of resting metabolic rate against body weight in 41 primate species. Data from Kleiber (1932).

The link between size and metabolic rate places basic constraints on an organism's life-history strategy to cover maintenance (with its little variability) but mostly growth and reproduction as adjusted variables. It is now generally agreed that energy availability supports growth before being used to support reproduction, as growth ceases at reproductive maturity (Charnov 1991), hence the strong heterochrony between body height, weight, and BMI, and both growth rate and reproductive human biology.

Total daily energy requirements of a healthy organism are determined by allometric factors such as body size—a correlate of the resting metabolic requirements activity, growth, and reproductive status. Over the course of hominid evolution, all four of these parameters experienced changes. Other than body size that has increased markedly (Fig. 10.1), activity costs, particularly those associated with foraging, are likely to have changed as our ancestors began to exploit their environment in new ways. Moreover, as discussed throughout this text, the evolution of our species prolonged the growth period and distinctive reproductive pattern.

Life-history theory is, at the end of the day, based on the law of thermodynamics, which states that energy used for one purpose cannot be used for another; thus, rates of energy harvest influence outcomes (Hill 1993). Life-history trade-offs result from limitations in the availability of critical resources such as energy or nutrients, necessitating choices on the differential allocation of resources to costly traits. For example, organisms that invest their resources primarily in reproductive function have fewer resources to invest in self-maintenance processes such as fat storage or immunity (Hau 2007). In the same way, adiposity rebound in the transition from childhood to juvenility is linked to decelerating longitudinal growth, and growth hormone deficiency to obesity.

Diet is directly related to energy harvest, and body size is also firmly intertwined with life history because it is critical to energy budget; at any given body size, a species that extracts higher-quality foods from the environment has increased life-history options for evolutionary fitness. On the other hand, while requiring more total energy, increasing body size improves heat retention and lessens the relative energy needed per gram of tissue (Schmidt-Nielsen 1975). An obvious trade-off for the metabolic cost, large body size decreases mortality, because there are always fewer predators for large mammals than for small ones.

Assuming similar activity budgets for all early hominid species, the estimated total energy expenditure for *Homo erectus* was 40–45% greater than for the *Australopithecus* (80–85% greater by other methodologies) (Leonard and Robertson 1997). The expansion of African savannas from 2.5 million to 1.5 million years ago might have been the impetus for a shift in foraging behavior among early members of the genus *Homo*. Such ecological changes would likely have made animal foods a more attractive resource. Moreover, greater use of animal foods and the resulting higher-quality diet would have been important for supporting the larger day ranges of mobility and greater energy requirements that appear to have been associated with the evolution of a human-like hunting and gathering strategy.

Evolution of life-history stages has involved the gradual cumulative selection and development of traits over millions of years. As a major consumer of energy, growth plays a major role, and the ability to vary growth patterns and the rate of growth during particular life stages, to shorten or prolong life stages, and even to add or delete life stages, provided a compelling strategy in adapting to the energetic cues at each evolutionary phase. The latter is exemplified by the addition of a childhood stage by *Homo habilis* 1.9 million years ago.

In apes, many of the ontogenetic variations can be understood with respect to Janson and van Schaik's model that predicted negative correlation between growth rates and ecological risks (Janson and van Schaik 1993). Specifically, this model anticipates the evolution of low growth rates when high levels of an ecological risk result from intraspecific feeding competition. Low growth rates minimize metabolic costs per time unit and can evolve in primates because selection for rapid maturation via predation tends to be minimized by group living.

Compared to our closest living and extinct relatives, humans have a large, specialized, and complex brain embedded in a uniquely shaped braincase. The evolution of large human brain size has had important implications for our energy balance (Leonard, Snodgrass et al. 2007).

From Darwin's *The Descent of Man*:

> Man in the rudest state in which he now exists is the most dominant animal that has ever appeared on this earth. He has spread more widely than any other highly organised form: and all others have yielded before him. He manifestly owes this immense superiority to his intellectual faculties, to his social habits, which lead him to aid and defend his fellows, and to his corporeal structure.

Large brains are energetically expensive, and human infants expend a larger proportion of their energy budget on brain metabolism than do other primates. It has been estimated that during human infancy, 87% of the resting metabolic rate is devoted to brain growth and function, as compared to 45% in the chimpanzee's infantile brain. By age 5, 44% of the resting metabolic rate is devoted to the brain (20% in the chimp), and the human adult brain consumes 20–25% (9% in chimpanzees) of its total resources. These high costs of large human brains are supported in part by our energy- and nutrient-rich diet. Indeed, among primates, relative brain size correlates positively with dietary quality, and humans fall at the upper end of this correlation (Leonard and Robertson 1997). Consistent with an adaptation to a high-quality diet, humans have relatively small gastrointestinal tracts, and an under-muscled but overly fat habitus as compared with other primates, features that help offset the high-energy demands of our brains.

Hominids, like other species, have developed mechanisms to adapt to feeding crises, which increased in frequency ever since *Homo sapiens* adopted his agrarian life-style and became susceptible to droughts. Some of these mechanisms are directly concerned with life-history stages. Whereas other mammalian species and nonhu-man primates continue lactation until the eruption of the first permanent molar teeth and then move directly into juvenility (self-foraging for food and caring for themselves), early cessation of lactation in humans required a trade-off. The burden is on the weaned infant; while he makes the transition into the newly evolved child-hood stage and maintains high metabolic demands for a brain that is still growing rapidly, he has had to evolve mechanisms to ascertain energy provision. While the typical 3-year-old is motorically too immature to forage or prepare food, and is limited by deciduous dentition, the human solution has been for older siblings, the extended family, and older members of the social group to provide food for children until they reach juvenility. These energetic considerations and trade-offs required that evolution leave room for plasticity to adapt to periods of energy crises or abundance.

The transitions from one life-history stage to the next correlate strongly with body size. Thus, the infancy–childhood transition is earlier in heavier children, as is the transition into juvenility/adrenarche (Cano, de Oya et al. 2006), and that into adolescence (Parent, Teilmann et al. 2003).

A. ENDOCRINE CONTROL OF ENERGY EXPENDITURE

The hypothalamus plays a key role in energy and weight homeostasis, but it remains a matter of debate how the hypothalamus, as the plasticity center, quantifies and conveys the environmental energy economy (Hauner and Hochberg 2002). A vast array of hormones have been implicated in presenting the energetic message to the brain: Leptin, as a fat-derived signal, certainly has some impact on all three transi-tion periods of adaptive plasticity (Hileman, Pierroz et al. 2000). Children with a positive energy balance, and therefore heavier, develop the juvenility transition-typical adiposity rebound earlier than lighter children (Fig. 4.D.1).

The same is true for insulin-like growth factor-I as an essential factor in the timing of childhood (Leger, Oury et al. 1996), juvenility (Baquedano, Berensztein et al. 2005) and adolescence transitions (Dunger, Ahmed et al. 2006). Whereas insulin-like growth factor-I has been discovered to be the growth hormone-mediating hormone, its level is a sensitive measure of the nutritional state, providing a possible mecha-nism for all energy-related life-stage transitions.

Another potential mediator of the signal is the neurosteroid DHEAS, which during female (but not male) juvenility correlates with leptin levels, suggesting an association between body fat and transition into juvenility/adrenarche (Blogowska, Rzepka-Gorska et al. 2005).

In that respect, leptin is more that a satiety hormone. The observation that obesity protects from osteoporosis suggested that the same hormones could regulate energy metabolism, bone mass, and growth, providing the impetus for the negative relationship of growth and adiposity. Testing this hypothesis revealed that leptin regulates bone mass through a hypothalamic relay, using two neural mediators, the sympathetic tone and CART (cocaine and amphetamine-mediated transcript), both acting on the same cell type: the osteoblast (Karsenty 2006). Leptin also has a direct effect on the growth plate, which has been shown to reverse the inhibitory effect of caloric restriction on longitudinal growth (Gat-Yablonski, Ben-Ari et al. 2004), suggesting that it might play a central role in conveying the energy cue to child growth.

The stomach and hypothalamus hormone ghrelin was found capable of stimulating pulsatile GnRH secretion from prepubertal rat hypothalamic explants, and has been implicated as a candidate messenger between energy balance and the fertility (Horvath, Diano et al. 2001).

Glucose and insulin are obvious signal messengers capable of modulating the reproductive axis. Their physiological role in terms of energy communication to the brain is complex and beyond the scope of this text.

B. WEANING AND GROWTH IN A MALNOURISHED ENVIRONMENT

A threat of famine has been the major evolutionary force in terms of energy economy, selecting for thrifty physiologic, metabolic, adipogenic gluttony (gorging while food is available), or behavioral genes. Yet, modern human children are exposed to oversatiety, as formulated by Neel in his "thrifty gene theory" (Neel 1962). Such maladaptions are handled within these periods of plasticity provided by the genome.

Protein-energy malnutrition is a syndrome resulting from interaction between poor diets and diseases, leading to anthropometric deficits and generally with deficits in micronutrients as well. While some may label it stunting, the growth deceleration of malnutrition, mostly in the form of delayed infancy–childhood transition, may represent an appropriate adaptive response to the energy circumstances (Hochberg and Albertsson-Wikland 2008). Figure 10.B.1 shows the much delayed infancy–childhood transition in malnourished children of Malawi that would result in an adaptive smaller adult stature (Z. Hochberg and K.E. Astrom, unpublished data).

Three anthropometric indices are commonly used as indicators of malnutrition: weight-for-age (underweight), height-for-age (stunting), and weight-for-height (wasting). A deficit SDS score (below −2) in any one of these indices reflects malnutrition, and a SDS score below −3 reflects a severe form of that condition. In developing countries, an estimated 50 million children aged younger than 5 years are malnourished, and those who are severely malnourished with a severe illness leading to hospitalization face a case-fatality rate exceeding 20% (Faruque, Ahmed et al. 2008). Based on earlier studies demonstrating that exclusive breast-feeding reduces morbidity and deaths from common infectious diseases, the WHO estimated that promotion of exclusive breast-feeding for the first 6 months could avert the deaths of 1.3 million infants globally each year (Black, Morris et al. 2003). Unlike the long breast-feeding in nonindustrial tribal societies, the situation for children in developing countries is by far less optimal. Although the prevalence of breast-feeding is very high in a country like Bangladesh, a fine representative for

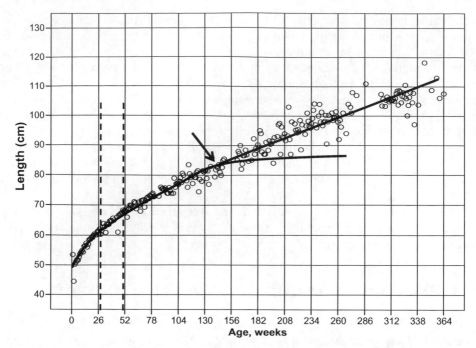

Fig. 10.B.1. The infancy–childhood transition in malnourished children of Malawi showing the much delayed transition age as compared to Swedish children (dashed line) that would result in a smaller adult stature. Data by Per Ashtrom.

developing countries, appropriate breast-feeding is rarely practiced. Infants are introduced to other foods either too early or too late; many infants remain on exclusive breast-feeding for 1–2 months only, when they are introduced to plain water and breast milk substitutes such as cow's milk or semi-solid gruels. Rather than transiting to childhood and accelerated growth, children in Bangladesh stunt their growth in response to complementary foods. Their transition to childhood at age 15 months onward reflects the unfavorable energetic cue, leading to an adaptive relatively short adult stature, whereas after transition to childhood, no further deterioration has been observed (Shrimpton, Victora et al. 2001).

11

STAGE TRANSITIONS: TRADE-OFFS AND ADAPTIVE PHENOTYPIC PLASTICITY

Natural selection acts on changeable traits of organisms. Such traits appear, and increase or decrease in frequency when they maximize long-term reproductive fitness, with fertility being the shortest link to an organism's fitness. Other fitness components, such as survivorship and mortality and interbirth intervals, influence fitness through their effects on fertility.

The two important trade-offs affecting fitness are between the present against future reproduction, and the trade-off between quantity and quality of offspring (Kaplan and Lancaster 2003). The trade-off for present against future reproduction is exemplified by boys' slow accumulation of somatic mass and height prior to sexual maturation, compromising the present and increasing their future fertility. The trade-off between present and future reproduction is also expressed in females, where investment in present reproduction, both through pregnancy and lactation, suppresses ovulation. In endocrine terms, progesterone generated by the placenta during pregnancy and prolactin secreted by the pituitary gland during lactation suppress the hypothalamic–pituitary–gonadal axis and prevent reproduction. Humans, like several other organisms, engage in repeated episodes of reproduction separated by the interbirth intervals, when the mother breastfeeds her infant. During the nonreproductive years, energy is diverted to maintenance and storage so that she can live to reproduce again. The trade-off between quantity and quality of offspring is a function of parental investment in offspring (high in humans) and affects their ability to survive to the next reproductive episode.

Evo-Devo of Child Growth: Treatise on Child Growth and Human Evolution, First Edition.
Ze'ev Hochberg.
© 2012 Wiley-Blackwell. Published 2012 by John Wiley & Sons, Inc.

Organisms' body size is a classic example of the combination of phenotypic adaptive plasticity and genetic evolution; a trait that is utilized by all orgranisms. The *Daphnia* (marine plankton known as water fleas) responds to a cue of fish predators by rapidly decreasing its size and enhancing its number of eggs (Latta, Bakelar et al. 2007). *Cnidaria medusa* (commonly known as jellyfish), whose propulsion requires the force of the bell contraction to generate forward thrust, develops a small body size in the high-thrust regime, and low fineness ratios and a large body size in the high-efficiency regime (Dabiri, Colin et al. 2007).

Another primary life function that is energetically costly in maintenance efforts is the immune system. Immune processes play a central role in cellular renewal and repair, and in defending against the damaging and potentially life-threatening effects of pathogenic agents; the body's investment in them supersedes at times all other considerations. Resources consumed by immune processes are not available to support investments in child growth or reproduction. The costs to growth is particularly high during the ultimate adult height-determining transition from infancy to childhood, and even more so in those infants with low-energy reserves (in the form of body fat) at the time of immunostimulation (McDade, Reyes-Garcia et al. 2008). As previously mentioned, insult at the infancy–childhood transition have lasting effects on child growth.

Whereas the terminology of phenotypic adaptive plasticity with respect to evolution is new, the concept is not. Charles Darwin observed in the first edition (1859) of *On the Origin of Species by Means of Natural Selection*, in the chapter "Struggle for Existence":

> We do not always bear in mind, that though food may be now superabundant, it is not so at all seasons of each recurring year.... A large number of eggs is of some importance to those species, which depend on a rapidly fluctuating amount of food, for it allows them rapidly to increase in number.

This is probably the first hint at adaptive plasticity in adjusting to changing environments.

A great deal of attention has been paid to intrauterine nutritional "fetal programming" leading to a "thrifty phenotype" and later health consequences (Wells 2009). In terms of fetal programming for adult height, intrauterine growth has minimal impact; the vast majority of infants with intrauterine growth retardation catch up immediately after birth and end up with a normal stature. Hence, in evolutionary terms, fetal growth, which depends on uterine and placental function, is too labile a stage to program for adult height.

Unlike the association of tall stature and longevity over evolutionary times, with humans being the longest living and the tallest among the primates, a general intraspecific trade-off exists between stature and survivorship. This may be linked to the growth hormone–insulin-like growth factor axis activity, which is positively related to size, but negatively related to longevity (Chen, Wu et al. 2010). In experimental animals, early life growth hormone treatment, which may parallel the human growth hormone-dependent childhood growth, shortened longevity and decreased cellular stress resistance (Panici, Harper et al. 2010). Findings based on millions of death reports suggest that shorter, smaller bodies have lower death rates and fewer diet-related chronic diseases, especially past middle age. Shorter people appear to have longer average life spans, even though low serum insulin-like growth factor-I is associated with increased risk of ischemic heart disease (Juul, Scheike et al. 2002). Animal experiments show that smaller animals within the same species generally live longer (Samaras, Elrick et al. 2003). The authors suggest that the differences in longevity between the sexes is due to their height differences with men average about 8% taller than women and have a 7.9% lower life expectancy at birth. This stature longevity trade-off may be related to later onset of puberty and, therefore, the older age at first fertility, which seems to program for a longer life span.

Organisms, including humans, show a remarkable regulatory flexibility that allows them to flourish under different external conditions and to survive harsh environments. Recent studies of gene expression reveal strategies for keeping regulatory mechanisms promoting adaptation to both short- and long-term environmental cues (Lopez-Maury, Marguerat et al. 2008). The balance between energy-efficient growth and the ability to respond to fluctuating environments is a basic physiological challenge. In response to an energy crisis, a multitude of genes induce or repress their transcription; mostly these are transient expression changes and return after some time to new steady-state levels that are close to those before the crisis.

Periods of energy crises often persist beyond a single generation in a given population. In preliminary unpublished observations on twins, we observed that the delay in the infancy–childhood growth transition with consequential short stature is a familial trait that is transmitted from parents to their offspring independently of their genes (German, Peter et al. 2009). Indeed, a negative correlation was demonstrated between the age of the infancy–childhood growth transition and fathers' height (Fig. 4.E.6) (Xu, Wang et al. 2002), indicating that offspring of ancestors, who might have responded to energy crises in their infancy, have a relative delay in the infancy–childhood growth transition. This is much in line with a report showing that whereas Maya-American children are much taller and have longer legs on average than Maya children in Guatemala, Maya migrants are shorter, even when the children were born in an affluent host country (Bogin 2002; Smith, Bogin et al. 2003). By age 4, most of the difference has been established, indicating the early age of adult height programming: at the infancy-childhood transition.

Adolescence is another life-history stage of adaptive plasticity, with a wide range of ages at transiting. Similar to earlier growth stages, about half the variance in the timing of menarche is familial (Towne, Czerwinski et al. 2005), but despite a large body of research, we have limited evidence that it is genetically determined (Gorai, Tanaka et al. 2003). What seems to be genetically determined is the plasticity to adapt to the environment, with energetic considerations as the major cues. These manifest most dramatically in adopted children who have migrated from deprived to affluent countries. The permutation of prenatal and early-life deprivation with juvenile nutritional overindulgence is predicted, and indeed, it advances transition into adolescence.

Each period of adaptive plasticity during transitions between life-history stages was discussed independently. Yet, these stages influence each other in an intricate web of connections that relate to evolutionary trade-offs for fitness, as measured by lifelong advantages. Prenatal growth strongly influences the transition into childhood, and transition into juvenility and adolescence. Infantile and childhood growth, and the timing of adrenarche, are related to the transition into adolescence and the age at menarche (Parent, Teilmann et al. 2003).

A. TRANSGENERATIONAL INFLUENCES IN LIFE-STAGE TRANSITION

It is common knowledge that prematurity—an obvious change in transition from one life stage to another—is a familial trait, but genes associated with this trait are but few and with a relatively small impact. This transition period is not dealt with in this treatise. Likewise, transitions from infancy to childhood and from juvenility to adolescence are familial traits with evidence for a small genetic impact on the variation. I speculate that intraspecial gene sequence variations may not play a major role in life-history stage transitions. The genes express plasticity/changeability but not the plastic changes.

As previously shown, these transgenerational traits are adaptive responses to environmental circumstances that last longer than a single generation. In any event, modern lifestyles, with rapid changes in world climate and massive rapid population shifts over long distances might have changed the rules, leading to a mismatch between the environment and the phenotype or between several phenotypic phenomena of each life-stage package (Gluckman and Hanson 2005).

Low birth weight is common in developing countries. But what happens to fetal growth in migrant mothers? Despite improved health care, such women give birth to smaller infants (Drooger, Troe et al. 2005), demonstrating the transgenerational influence of their ancestral environment.

In a study in rural Guatemala, pregnant women and their offspring received one of two types of nutritional supplements. One was a drink called "atole," which was enhanced with protein and calories; the other was a drink called "fresco," which was enhanced with calories, but less than the calories in the atole drink. The infants and children of the original study received these nutritional supplements until age 7. Then, in the 1990s, women who were part of the original study as children were asked to be part of a follow-up analysis to investigate the transgenerational influences of the original intervention (Stein, Barnhart et al. 2004; Bogin, Silva et al. 2007). The mothers had been measured repeatedly from their birth until age 3, as were their children. The report shows that the current generation of infants grew in general faster than their mothers did (Fig. 11.A.1). The infancy–childhood transition occurred in the mothers' generation at a mean age of 20 months, but earlier in their sons and daughters. However, 30 years after the initial intervention, infants of women who received the protein-energy drink atole grew faster than infants of mothers who received fresco (calorie-rich only), showing that the rate of child growth reflects, in part, the growth pattern of the mother, including improvements to that pattern resulting from nutritional supplementation.

The global prevalence of obesity and its familial transmission is clear; having either parent obese is an independent risk factor for childhood obesity. A recent study showed nongenetic factors in the causal pathway: paternal high-fat-diet exposure programmed pancreatic beta-cell "dysfunction" in his female offspring (Ng, Lin et al. 2010). Chronic high-fat-diet consumption in fathers induced increased body weight, adiposity, impaired glucose tolerance, and insulin sensitivity. Relative to controls, the fathers' female offspring had an early onset of impaired insulin secretion and glucose tolerance that worsened with time, and normal adiposity. A paternal high-fat diet altered the expression of 642 pancreatic islet genes in adult female offspring. In the next chapter we will discuss epigenetic changes, but here, hypo-

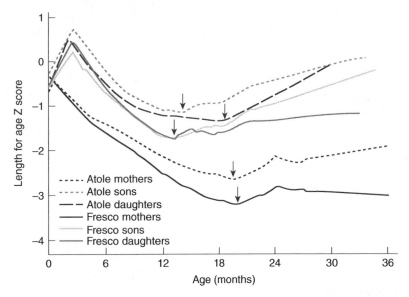

Fig. 11.A.1. Transgeneration impact of nutrition. Growth of women 1969–1977 and their children 1996–1999 in Guatemala in the first 3 years of life on higher- (atole) or lower-quality food (fresco), expressed as deviations from the World Health Organization reference population. Adapted with permission from Stein et al. (2004). The infancy–childhood transition occurred in the mothers' generation at a mean age of 20 months, but earlier in their sons and daughter.

methylation of the Il13ra2 gene showed the highest fold difference in expression (Ng, Lin et al. 2010).

B. EPIGENETICS AND LIFE-HISTORY STAGE TRANSITIONS

"Plasticity," "transgeneration influence," and "adaptive responses"—terms that have been widely used in this treatise—are also the jargon used in epigenetics, a 60-year-old discipline that has been making a comeback. There is no experimental evidence that any of the following are based on epigenetic mechanisms, yet the similarities are so obvious that one can hardly resist the speculation (Hochberg, Feil et al. 2011).

The current major paradigm for disease etiology includes classic genetics in which DNA sequence mutations are a major etiological factor. From the epidemiological studies of Forsdahl and Barker, it has now become evident that the environment influences disease and its prevalence (Forsdahl 1977; Barker and Osmond 1986), and many environmental factors and compounds influence disease prevalence. For example, individuals who move from one region in which there is particular disease prevalence to another region with different disease prevalence are more likely to develop diseases that are prevalent in the new location (Gluckman and Hanson 2007a).

More recently, epigenetics has been added as a source of influence on gene expression profiles and transcriptomes of most organs and cell types. Epigenetic modifications to DNA provide a plausible link between the environment and alterations in gene expression that might lead to disease phenotypes (Jirtle and Skinner 2007). There is an increasing body of evidence from animal studies that support the role of environmental epigenetics in disease susceptibility (Jirtle and Skinner 2007). Accordingly, alterations in the epigenome are now considered to be major factors in the regulation of the transcriptomes that are associated with disease (Egger, Liang et al. 2004). The ability of an environmental factor to promote a permanent epigenetic change in the germline suggests an epigenetic component in disease etiology, and a molecular mechanism for the ability of environmental factors to influence disease. Thus, the genome is programmed by the epigenome, and the epigenome is the genome's interface with the environment (Szyf, McGowan et al. 2008).

Epigenetic mechanisms result in stable regulation of gene expression without any alterations to the gene's DNA, and trigger initiation and/or maintenance of cell-specific transcriptional profiles. Indeed, the precise control of transcription is achieved by modulating the chromatin structure and three-dimensional organization of the genome and nuclear architecture. Epigenetic marks are flexible. Thus, it is possible for environmental factors that include nutrition, social cues, cultural background, hormones, drugs, and toxins to alter epigenetic landscapes in a specific spatiotemporal window and in a tissue- and sex-specific manner. Alterations of epigenetic marks can lead to irreversible changes in lineage specification. Therefore, early deviation of cell-type determination by either amplifying or decreasing the number of specific cell subtypes can lead to disease and/or changes in disease susceptibility in adulthood (Haumaitre, Lenoir et al. 2008), or to phenotypic changes, all of which could be reversed using appropriate epigenetic tools (Szyf, McGowan et al. 2008). In addition, links have been found between epigenetic alterations and (a) circadian, sleep-wake, and rest-activity rhythms; (b) the hunger-satiety cycles; and (c) the major components of energy homeostasis and thermogenesis (Gallou-Kabani, Vige et al. 2007).

Expression of genes in differentiated tissues is controlled by differential silencing of genes by epigenetic processes such as DNA methylation and histone modification and microRNA. Whereas the genome is obliterated of most epigenetic marks early in embryogenesis, tissue differentiation is associated with replication of the epigenetic marks, so that these marks are transmitted across generations. It was shown that environmental cues lead to stable changes in the epigenome, which provide a mechanistic platform for plasticity and transgenerational transmission of its impact.

There is increasing evidence that epigenetic events underlie developmental plasticity where the impact of environmental triggers on phenotypic variance that are felt in the initial generation are "transmitted," and therefore influence the development and behavior of subsequent generations (Hochberg, Feil et al. 2011). The epigenetic mechanisms are numerous, complex, and interact with each other, and can be modified during development. For example, there is considerable evidence for prenatal manipulation of DNA methylation by hormonal stimuli, and maternal nutritional can influence gene expression of the fetus (Wolff, Kodell et al. 1998; Waterland and Jirtle 2004; Dolinoy, Huang et al. 2007). Furthermore, epigenomic reversal is possible because plasticity extends beyond birth—with multiple examples in this book. Indeed, initial data gathered by experimental replication of the developmental origins of health and adult disease (DOHaD) phenomenon (Vickers, Gluckman et al. 2005; Burdge, Lillycrop et al. 2008) and its reversal in experimental animals adds further support for the likely role of epigenetic mechanisms in this form of developmental plasticity.

One mechanism by which maternal diet during pregnancy may lead to stable changes in gene expression in the offspring is by epigenetic regulation of genes. Fertile bee queens and sterile workers are alternative forms of the adult female honeybee that develop from genetically identical larvae following differential feeding with royal jelly. Silencing the expression of DNA methyltransferase Dnmt3, a key driver of epigenetic global reprogramming, in newly hatched larvae, led to a royal jelly-like effect on the larval developmental trajectory. The majority of Dnmt3 small interfering RNA-treated individuals emerged as queens with fully developed ovaries (Kucharski, Maleszka et al. 2008).

Coat color in mice is determined by the methylation status of the intracisternal A particle (IAP) retrotransposon at the 5' end of the agouti gene, and maternal diet can affect DNA methylation in the offspring. Supplementation of maternal diet with methyl donors shifted the coat color of their offspring to the pseudoagouti phenotype (Waterland and Jirtle 2003). Hypomethylation was associated with a yellow coat in the offspring, whereas hypermethylation was associated with a brown coat.

Challenges during pregnancy or neonatal life can result in changes in gene promoter methylation and affect gene expression in pathways that are associated with various physiological processes. For example, the role of epigenetic regulation of stable gene expression in the induction of an altered phenotype by maternal nutrition has been investigated by Lillycrop et al. in pregnant rats and mice that were fed either a protein-restricted diet or were exposed to undernutrition (Lillycrop, Phillips et al. 2005, 2008). The primary readouts of these experiments were the hepatic expressions of the glucocorticoid receptor (GR) and peroxisome proliferator-activated receptor-α (PPARα), as well as other markers of glucose homeostatsis and β-oxidation. They chose these two genes because their proteins are nuclear receptor proteins, and both play roles in metabolism. They showed that modest changes to maternal intake of macronutrients during pregnancy induce stable changes to the epigenetic regulation of GR and PPARα in the liver of juveniles and adult offspring (Burdge, Slater-Jefferies et al. 2007).

Collectively, these examples suggest that altered gene methylation during prenatal life may provide a causal mechanism to explain how maternal diet can induce stable changes in gene expression within offspring and may represent a fundamental mechanism for altering the phenotype.

Supporting this experimental work, a study of human umbilical cords suggested similar relationship in humans (Lillycrop, Slater-Jefferies et al. 2007). The umbilical cords were collected from 15 term infants whose birth weights were within the normal range, and they then determined the DNA methyltransferase mRNA expression in the cords in order to establish whether enzyme expression was related to the level of GR methylation in fetal human tissue. In addition to expression of the transcripts from the human GR promoter, the researchers reported that the methylation status of a gene expressed in human fetal tissue varied considerably for infants that were born within the normal birth weight range. In addition, the level of methylation correlated inversely with GR expression in the cord, and correlated positively with Dnmt 1 expression. Thus, methylation of the human GR is associated with the capacity of Dnmt1 to maintain methylation of CpG dinucleotides, rather than the capacity for DNA methylation *de novo*. These findings are consistent not only with their findings in rats, but are also consistent with their hypothesis that the induction of different phenotypes in humans by prenatal nutrition may involve variations in Dnmt1 expression and, in turn, DNA methylation. When all their findings are considered together, they lend support to the notion that that there are critical periods during embryogenesis and early postnatal life when epigenetic processes are susceptible to perturbations by maternal nutrition.

There are studies also that explored the "rescue" of aberrant phenotypes *in utero* or the reversibility of induced phenotypic effects. Lillycrop et al. showed that supplementation of the protein-restricted diet with folic acid, a methyl donor cofactor, during pregnancy prevented changes to the methylation status of the GR and PPARα promoters, and led to the normalization of GR and PPARα expression (Lillycrop, Phillips et al. 2005). Feeding either a folic acid- or glycine-enriched protein-restricted diet prevented an altered phenotype in the offspring (Jackson, Dunn et al. 2002; Brawley, Torrens et al. 2004). In addition, folic acid supplementation in the peripubertal period did not normalize the effect of the maternal protein-restricted diet on the phenotype or epigenotype of the offspring (Burdge, Lillycrop et al. 2009).

Instead, supplementing the diet of the offspring with folic acid during the peripubertal period increased their weight gain, but impaired their capacity to maintain lipid homeostasis, despite promoting DNA methylation. These findings imply that the transition from juvenility to adolescence is yet another period of plasticity when specific nutrient intake may alter the phenotype of the offspring through epigenetic changes in specific genes. This is by no means a recommendation to prescribe folic acid in transition periods; folic acid supplementation can cause other effects that may not be desired.

Such findings suggest that DNA methylation is used for storing epigenetic information. In addition, such findings demonstrate that the use of this information in the offspring in its early stages of development can be differentially altered by nutritional input, and that the flexibility of epigenetic modifications underpins profound shifts in developmental trajectories and fates.

Understanding the mechanisms that are responsible for such induced epigenetic changes can have implications for reproductive and behavioral status. In addition, understanding the mechanisms may provide novel biomarkers of risk and new opportunities for therapeutic strategies to prevent or ameliorate the effects of adverse events in the early life environment on disease risk (Hochberg, Feil et al. 2011).

Epigenetics refers to processes that stably alter gene activity without altering the gene sequence. As epigenetic processes are integral in determining when and where specific genes are expressed, alterations in the epigenetic regulation of genes may lead to changes in phenotype. The two major epigenetic mechanisms are DNA methylation and histone modification (Bird 2002). Methylation at the 5' position of cytosine in DNA within a CpG dinucleotide is a common modification in mammalian genomes, and constitutes a stable epigenetic mark that is transmitted through DNA replication and cell division. Methylation of CpG-rich clusters (CpG islands), which span the promoter regions of genes, is associated with transcriptional repression, while hypomethylation of CpG islands is associated with transcriptional activation (Bird 2002). These methylation patterns are established largely *in utero* or early postnatal life. Methylation of CpG dinucleotides *de novo* is catalyzed by DNA methyltransferases, and is maintained through mitosis by gene-specific methylation of hemimethylated DNA (Reik, Dean et al. 2001). However, once these methylation patterns have been established during development, these epigenetic markers are in most cases maintained with high fidelity throughout life. The periods during development when these methylation patterns are established have been shown to be susceptibility to early life environmental influences (Zeisel 2009).

Whereas the evolutionary origin of the switch and control mechanisms for onset of the childhood growth stage is genetic, it is speculated that the plasticity of the growth transition switches may be epigenetic, causing heritable changes in gene function without changing the DNA sequence. I concede the lack of any experimental epigenetic studies to support this speculation. However, epigenetic changes occurring at certain stages of development become incorporated into the genome, providing transgenerational passage of environmentally induced epigenetic changes (Anway, Cupp et al. 2005), and endocrine, physiological and behavioral traits persist even after the original selection pressure is relaxed (Sollars, Lu et al. 2003) for at least three generations (Anway, Cupp et al. 2005; Sollars, Lu et al. 2003; Campbell and Perkins 1988). Because such events can occur at a rate up to 100,000 times higher than point mutations, epigenetic changes potentially have a much greater impact on morphological adaptive phenotypes (Ellegren 2000).

In the case of a delayed infancy–childhood growth transition, DICT hormones, cytokines, and nutrition may account for epigenetic evolutionary adaptations to energy crises. Hormones have been shown to express from epigenetically imprinted genes, and hormone target genes are liable to be epigenetically modified. Hormonal exposure early in life alters the response to hormonally regulated genes to the same or different hormones later in life; the first hormone experience alters the set point epigenetically for a later hormone response, the so-called two-hit model. More specifically, exposure to estrogens early in life compromises health later in life, fetal glucocorticoids imprint later glucocorticoid responsiveness (Moritz, Boon et al. 2005), demethylation of the leptin promoter correlates with preadipocyte differentiation into adipocytes (Melzner, Scott et al. 2002), and leptin has a trophic effect on plasticity of hypothalamic responses to nutrients that regulate feeding (Bouret, Draper et al. 2004). As DICT is associated with delayed setting in of the growth hormone–insulin-like growth factor-I axis, it is plausible but has yet to be determined whether epigenetic mechanisms are involved at the level of hormone secretion or target cell responsiveness.

Nutritional deprivation during infancy, even if transient, may have a lasting influence on the expression of genes by interacting with epigenetic mechanisms and altering chromatic conformation and transcription factor accessibility (Waterland and Garza 1999). Moreover, epigenetic programming by energy crises has been shown to be transmitted to the next generation (Waterland and Jirtle 2004).

It was previously mentioned that a shortage of the food supply during the juvenility–adolescence transition had a positive effect on life span two generations later, and that it was transmitted on the paternal side from grandfathers to grandsons and from grandmothers to granddaughters (Pembrey, Bygren et al. 2006). Little is known about the transmitting mechanisms, except in a specific study of epigenetic state and the altered DNA methylation patterns at two loci in the epididymal sperm of rats with reduced spermatogenic capacity due to their paternal ancestor *in utero* exposure to endocrine disrupters (Anway, Cupp et al. 2005).

In the previous chapters, I discussed transgeneration transmission of an adaptive phenotype. Some epigenetic marks may originate from a previous generational experience (Jirtle and Skinner 2007). Epigenetic marks that failed to be erased before implantation or in the germline may be transmitted to the next generation in a sex-specific manner and exert a transgenerational effect (Dasenbrock, Tillmann et al. 2005; Whitelaw and Whitelaw 2006). During critical periods of life (periconception, fetal, and life history phase transitions), exposure to deleterious environmental compounds, abnormal maternal behavior, or inadequate maternal feeding can induce various developmental alterations. Such alterations can increase disease susceptibility to the offspring that can also be transgenerational, because disease susceptibility can be transmitted to subsequent generations. The results of early studies on transgeneration effects assumed that they were the result of the epigenetic malprogramming of somatic processes. However, paternal or maternal germline epigenetic inheritance may also account for the transgenerational effect (Anway, Cupp et al. 2005). Moreover, both somatic and germline effects may be sexually dimorphic, and can affect both mitochondrial and the nuclear DNA through the maternal line (Taylor and Denardo 2005).

C. NOTE BY KEN ONG ON POPULATION GENETICS AND CHILD GROWTH AND MATURATION

1. Genetic Adaptation

Humans adapt to their environments through a wide variety of mechanisms, including metabolic-endocrine, epigenetic, and genetic processes. While these adaptations optimize fitness, the mismatch hypothesis states that these processes may lead to disease risk, particularly if the environment subsequently changes or is not that which was originally expected.

Take the example of adaptation to hypoxia. Following the almost immediate physiological rise in heart rate and ventilation rate, metabolic changes over a period of hours or days increase the renal excretion of bicarbonate to prevent the adverse effects of respiratory alkalosis. While other metabolic changes occur, the main endocrine response is the rise in erythropoetin secretion leading to haematological adaptation, an increase in red blood cell number, over a period of several weeks (Martin, Levett et al. 2010). While this classical adaption is associated with the adverse risks of polycythemia, a recent study identified a novel genetic adaptation to chronic hypoxia among high altitude dwellers in Tibet who show profound arterial hypoxia and yet maintain normal aerobic metabolism (Simonson, Yang et al. 2010). By comparison of genome-wide single nucleotide polymorphism (SNP) frequencies in Tibetans to those in neighboring lowland populations, this study identified evidence for evolutionary selection of several genes with *a priori* functional roles in hypoxia response. Remarkably, each additional copy of the identified hemoglobin concentration, which is consistent with the suggestion that high-altitude Tibetans have evolved more efficient oxygen transport systems of adaptation to hypoxia (Simonson, Yang et al. 2010).

Therefore, although they operate on a far longer timescale than those other processes, genetic adaptations may increase fitness in certain environments. In different settings these adaptations may increase the risk of adverse health. In the example above it is unclear how adaptive oxygen transport systems may relate to disease risk, but other well-known examples of genetic traits with major survival benefits (for example, sickle cell and alpha thalassemia traits in relation to malaria infection and disease) also confer susceptibility to adverse hematological events. Genetic variants that confer lactase persistence are thought to have shown positive selection since the expansion of agricultural societies (Harris and Meyer 2006). In recent years there have been reports of possible genetic adaptations that impact on childhood growth and development. These will be discussed below.

2. The Genetic Epidemiology of Child Growth and Maturation

With regard to Hochberg's premise that evolutionary development biology explains the regulation of childhood growth and maturation and related disease, genetic factors contribute a substantial proportion of the wide variation in these childhood traits and may provide important clues both for the specific processes involved and how adaption may have arisen.

3. Basic Principles and Heritability Estimates from Twin Studies

The study of twins has long been used to help distinguish between the contributions of genes and environment, which is loosely defined as all nongenetic factors. The basic design relies on the distinct genetic characteristics of monozygotic as compared to dizygotic twins. Monozygotic twins arise from a single fertilized ovum and are therefore genetically identical. Dizygotic twins originate from two separate ova and share on average half of their genes. The heritability of a trait or disease, defined as the proportion of the total variance that is explained by genetic factors, can be derived from the comparison of the intrapair similarities between monozygotic vs. dizygotic twins. For many childhood growth characteristics, the pattern of intrapair correlations tends to be different, and greater similarities between monozygotic, as opposed to dizygotic pairs, point to genetic influences. In addition to simple heritability calculations, structural equation modeling, allows in-depth insights into the genetic and environmental architecture underlying variation in disease or traits.

Studies of the family history of obesity, adoption, and twin studies of obesity demonstrate that individual susceptibility to obesity has a strong genetic component (Stunkard, Sorensen et al. 1986; Sorensen, Holst et al. 1992b; Parsons, Power et al. 1999). Such observations have underpinned large-scale efforts to identify the common genetic determinants of obesity. In a recent systematic review of childhood twin studies, genetic factors appeared to have a strong effect on the variation of BMI during childhood with estimates ranging between 60% and 90% (Silventoinen, Rokholm et al. 2010). Other studies conclude that the genetic influence on body mass index at age 7 years is as strong as that in adult life (Sorensen, Holst et al. 1992a). In addition to childhood obesity, studies of twins and extended families support a strong genetic component to childhood height (Tanner 1985). For the timing of puberty, studies estimate that around 50–70% of the variance in the age at menarche in girls may be attributable to genetic factors (Towne, Czerwinski et al. 2005; van den Berg and Boomsma 2007).

4. More Complex Heritability Models

In addition to cross-sectional body size, studies of twins have modeled longitudinal changes in childhood growth. Studies of infants and young children report strong genetic components for infancy and childhood weight gain ranging from 56% to 82% of the variance (Demerath 2007; Beardsall, Ong et al. 2009), and also 67–78% for patterns of growth in infant recumbent length (Towne, Guo et al.1993). Growth data have also been modeled to estimate the timing of puberty in boys, where the lack of noninvasive measures of pubertal timing has lead to a scarcity of studies. A simple measure HD:SDS (height difference in SD scores) was calculated as the difference in height SDs between 14.0 and 17.5 years of age. This was validated against more detailed measures of the timing of peak height velocity and reported pubertal development, and was used to estimate consistently high heritabilities of pubertal timing of 86% and 82% in girls and boys, respectively.

Complex statistical genetic modeling can be used to estimate the changes in heritability with age and also the coinheritance of related phenotypic traits. For example, Silventoinen, Bartels et al. (2007) applied a longitudinal twin model to estimate the heritability of growth between the ages of 3 and 12 in 7,755 twin pairs. With increasing age the heritability estimates decreased and environmental factors become more important for both height and BMI. However, in other longitudinal twin studies the heritability of BMI has been shown to increase over childhood (Haworth, Carnell et al. 2008) and decrease with age in adults (Korkeila, Kaprio et al. 1991), which is consistent with the results of an emerging wave of studies using specific genetic factors (Hardy, Wills et al. 2010).

Studies using models with multiple outcome variables have suggested that there is a substantial coinheritance of the timing of puberty and BMI. In a bivariate twin analysis of age at menarche and BMI, Kaprio, Rimpela et al. (1995) estimated that 37% of the variance in age at menarche was attributable to additive genetic effects, a further 37% to dominant genetic factors, and only 26% to unique environmental factors. Of note, the correlation between additive genetic effects on age at menarche and BMI was 0.57, indicating a substantial proportion of genetic effects in common. This prediction was subsequently to be proved accurate with the identification of many specific genetic factors in common to both BMI and the timing of menarche (Elks, Perry et al. 2010).

5. Heritability Is Dependent Upon the Setting

The classical twin study design is dependent on a number of assumptions and limitations, such as assuming that twins are representative of the general population, whereas twins are around 600 g lighter than singletons and catch up rapidly within the first few months of life (Wilson 1979). The major assumption that the shared environment between members of a pair is the same for monozygotic and dizygotic twins has been questioned (Richardson and Norgate 2005) and may lead to an artificial inflation of heritability estimates. Over-optimistic heritability estimates may also arise from the failure to capture important contributions of environmental factors that act over different settings and eras. Examples are the marked secular trends in Western settings to the eradication of childhood undernutrition and the rise in obesity.

As most twin studies are set in modern Western settings, it is possible to view the scientific literature as being biased towards overestimation of heritability. However, it is important to recognize that there is no "true single value" for the heritability of any trait. Rather, the impact of genetic factors on overall phenotypic variation depends on the prevailing environment. Hence, among an invariable population of smokers, lung cancer would appear to be almost completely heritable. In contrast, among large kindreds with mutations in the hepatic enzyme phenylalanine hydroxylase, the congenital metabolic disease phenylketonuria might appear to be a nutritional disorder related to intake of meat, fish, and various dairy products. All we can say is that in contemporary Western settings, wide variations in childhood growth and maturation are still apparent and these appear to be substantially genetic traits.

6. Essential Genes for Childhood Growth and Maturation

Detailed observation and characterization of rare human cases with deleterious mutations have established that the endocrine pathways regulate both childhood growth and pubertal development. Defects in genes encoding growth hormone (GH), the GH receptor (GHR), insulin-like growth factor-I (IGF-I), the IGF binding proteins and receptors, and more recently the IGF signaling cascade, all underlie rare cases of severe short stature (Rosenfeld 2006). With regard to sexual maturation, identification of rare defects in neurokinin B and kisspeptin signaling are helping to delineate the hypothalamic signaling of the pituitary–gonadal axis.

While these genes, and the proteins they encode, may be regarded as essential for normal childhood growth and maturation, for the most part the identified deleterious mutations occur far too rarely to explain the wide normal varieties that are characteristic of childhood growth and maturation even within settings of adequate nutrition.

Rather, heritability is likely to be explained by infrequent or even common polymorphisms with more modest influences on gene expression, protein levels, or function. In recent years, much progress has been made in identifying common variants for these childhood traits, and these findings have lead to new insights into its biological mechanisms and possible evolutionary influences.

7. Common Genetic Variants for Childhood Growth and Maturation

Until a few years ago, the search for common genetic variants related to common diseases required a gene-by-gene search whereby common variants in biologically plausible candidate genes were tested for association with disease. While some positive findings were replicated in other studies, in retrospect the vast majority have proved to be false positives. For pubertal timing, candidate gene approaches have not yet identified specific common genetic variants that convincingly influence the timing of puberty (Gajdos, Butler et al. 2008). The candidate gene approach is limited by our narrow understanding of the biological basis of these disorders and also by our yet poor understanding of which areas of the genome regulate gene function. In contrast, the "hypothesis-free" "genome-wide association" study (GWAS) approach has yielded many rewards in the study of human disease genetics (Hindorff, Junkins et al.).

TABLE 11.1. Cumulative Number of Common Genetic Variants *Robustly** Associated With Measures of Growth and Maturation

	Year				
	2006	2007	2008	2009	2010
BMI	0	1 *FTO*	2 *+MC4R*	12	30
Height	0	1 *HMGA2*	47	47	180
Age at menarche	0	0	0	2 *LIN28B* & 9q31.2	32

*Defined as reaching genome-wide statistical significance ($P < 5 \times 10^{-8}$) ± evidence of replication in other studies.

GWA studies genotype hundreds of thousands of SNPs spread across the entire genome. This technology requires large case control or population studies to provide sufficient statistical power to offset the hundreds of thousands of tests involved, and such large studies are therefore often dependent on collaborative efforts between several groups.

In recent years, GWAS papers have reported an exponential rise in the number of specific genetic loci that are robustly associated with anthropometric traits or timing of maturation (Table 11.1). The word "robustly" can be interpreted as both reaching genome-wide statistical significance, that is to say corrected for all common variants throughout the genome, and also confirmation or replication in large parallel and/or subsequent studies. Such strict criteria are critical to ensuring the "filtering-in" of findings that are true positives. For each trait, the initial papers invariably reported only one or two loci. These are the "lowest-hanging fruit," that is, those SNPs with the largest contributions in terms of effect size and minor allele frequency (MAF). Subsequent studies are much larger in size and consequently have much greater power to detect SNPs with more modest effect sizes and/or smaller MAFs.

Despite the astonishing pace of recent progress, the identified loci currently explain only around 2–5% of the heritability of these traits; so much remains to be discovered (Manolio, Collins et al. 2009). The current international collaborative approach of pooling summary association statistics from population studies will continue to grow as new studies assay their samples using an increasing variety of high-throughput genotyping chips. In addition, the next generation of population genetics studies is set to grapple with the enormous task of analyzing mountainous volumes of whole genome sequencing data.

8. GWAS Findings Lead to New Biology

One notable observation on the initial GWAS findings for adult height and age at menarche was the absence of signals representing the traditional endocrine pathways. In the very latest high-yielding GWA studies for adult height, it is reassuring to see associated variants in or at least near to most of these expected genes, such as *GH, GHR, IGFI,* and *IGFRI* (Allen, Estrada et al. 2010). However, the vast majority of loci with larger effects represent yet unknown regulatory pathways and point scientists toward new biological understanding of these traits.

A number of loci for adult height are related to microRNAs, particularly those of the *let-7* family of microRNAs, and their functional processing. Furthermore, one of the height loci, at *LIN28B* was identified as one of the first two loci for age at menarche and was shown to be relevant for the timing of puberty in both boys and girls. *LIN28B* is homologous in sequence to the heterochromic gene *lin-28* in *Caenorhabditis elegans* (Guo, Chen et al. 2006). Deleterious mutations in *lin-28* produce an abnormal rapid tempo of development through larval stages to adult cuticle development (Ambros and Horvitz 1984). Conversely, enhancement of *lin-28* expression delays larval progression (Moss, Lee et al. 1997). In humans and other mammals, *LIN28B* and its homolog *LIN28A* encode potent and specific regulators of preprocessing of the *let-7* family of microRNAs (Viswanathan, Daley et al. 2008), and regulate cell pluripotency (Yu, Vodyanik et al. 2007).

Subsequent to the GWAS reports, Zhu et al. found that genetic modification of the *Lin28-Let-7* pathway in mice by over-expression of *Lin28a* leads to increased body size and delayed onset of puberty (Zhu, Shah et al. 2010). *Lin28a* transgenic mice were heavier and longer than wild-type mice after weaning and had a pro-longed growth period. The mechanism of rapid growth involved cell hyperplasia rather than hypertrophy, and also increased insulin sensitivity and glucose uptake. These findings illustrate how the whole genome association approach can lead to new directions in our understanding of human biology. They suggest a remarkable conservation from worm to human of a fundamental cell regulatory system that controls the tempo of somatic development, and also suggest a physiological role for microRNA processing in the timing of human growth and development. A later collaborative GWAS for menarche timing (Elks, Perry et al. 2010) reported an overlap between loci for BMI and menarche timing, in keeping with the coinheritance of these traits predicted by twin studies (Kaprio, Rimpela et al. 1995). It also highlighted genes in the fatty acid biosynthesis pathway as a possible novel mechanism to explain the signaling of nutrition on reproductive maturation and function.

9. GWAS Findings Lead to New Phenotypic Understanding

In addition to new molecular understanding, there is a growing awareness that these genetic findings, though yet far from complete, can aid our understanding of other phenotypes that are related to the trait of interest. For example, the menarche timing variant in *LIN28B* was particularly strongly associated with the timing of voice breaking in boys, and this finding highlights that trait as a robust and noninvasive measure of pubertal timing in boys. Secondly, statistical causal modeling, the so-called mendelian randomization studies, allow epidemiologists to make inferences from purely observational data regarding the causality and causal direction of association between phenotypic traits.

In a contemporary UK birth cohort study, Elks et al. reported that several of the GWA studies identified that genetic variants for adult BMI have a combined association with childhood weight gain that is apparent even within the first weeks after birth (Elks, Loos et al. 2010). Rapid infancy weight gain and larger infant body weight have been consistently related to increased risk of obesity in subsequent childhood or adult life (Baird, Fisher et al. 2005; Ong and Loos 2006). However, with regard to hard disease outcomes, studies in historical cohort studies have been conflicting. For example, rapid infant weight gain has been associated with increased risk of obesity and the metabolic syndrome, but with reduced risk of type 2 diabetes (T2DM) (Fall, Sachdev et al. 2008). One difficulty is that studies in historical cohorts or in societies that are undergoing nutritional transition may identify life-course associations that are specific to those particular settings (Stettler 2007). Even in contemporary Western studies, the very long-term follow-up needed to record adult disease outcomes may mean that current findings will not be applicable to future infant settings. The application in contemporary birth cohorts of genetic markers that are robustly associated with adult disease risks may provide a novel approach to life-course epidemiology by identifying early exposures that are directly relevant to current settings.

In their study, Elks et al. also reported that even in a contemporary affluent Western populations, the "adult obesity risk loci" appeared to have just as much effect on preventing slow weight gain as on promoting rapid weight gain and the risk of overweight (Elks, Loos et al. 2010). That remarkable observation may give us a clue as the physiological role of these common variants as, in view of the historical timeline of genetic adaption discussed above, we can be sure that these variants did not evolve for the promotion or protection of obesity per se. Failure to thrive in infancy, variably defined as underweight or poor weight gain, is a multifactorial condition with high mortality rates in historical settings (Jolley 2003). After exclusion of certain organic causes in a minority of cases, the majority of infants with failure to thrive likely represent the lower-normal distribution of infancy weight gain. Health professionals have increasingly recognized the contribution of innate variation in infant food demand rather than simply food provision by the caregiver (Skuse 1985). Protection against poor infant weight gain may represent a potential advantage associated with genetic obesity susceptibility alleles in some settings.

10. Genetic Adaptations for Childhood Growth and Maturation

James Neel proposed the "thrifty genotype" hypothesis to explain the extraordinarily high prevalence of obesity and type 2 diabetes among native Pacific Islanders who had experienced a sudden transition from chronic nutritional scarcity and intermittent famine to nutritional abundance with the advent of phosphate mining and Western food programs in the 1900s (Neel 1962). Neel purported that these populations had adapted genetically to survive and reproduce effectively in those earlier markedly adverse conditions by somehow being able to more efficiently store energy as fat whenever food was available. With the abrupt change in energy availability, they were suddenly exposed as being metabolically naïve to the different nutritional challenges. Yet since that proposal over 50 years ago, and an accompanying evolution in the theorized mechanism of metabolic adaption, there have been no specific genetic candidates identified to explain the adaptation to energy scarcity (Speakman 2008).

Infancy and early childhood weight gain fits the criteria for a trait that is under genetic adaption pressure, and probably much more so than adult fat storage, in view of the shockingly high rates of mortality during this period of life in pre-

Western settings. In some developing societies, it is estimated that only half of newborn children survive to the age of 5 years old. The majority of deaths occur in the first few weeks of life, when establishment of adequate breast milk supply in response to the stimulus of infant suckling is crucial. Subsequently, much of the mortality is attributable to infectious disease where nutritional status may have a major impact on immune function.

Another potential childhood trait that these obesity variants influence is the timing of sexual maturation. Prentice has argued that reproductive traits are the natural targets for genetic adaption that confer greater reproductive fitness during energy scarcity but predispose to obesity during more affluent settings (Prentice, Hennig et al. 2008). It is therefore notable that adult BMI loci were invariably found to be also associated with earlier age at menarche (Elks, Perry et al. 2010), a trait that may be reproductively advantageous in highly adverse settings but is associated with greater risks of obesity, type 2 diabetes, and mortality in Western settings (Lakshman, Forouhi et al. 2009).

There is as of yet no evidence for genetic adaptation at these loci. This will require detailed study in populations exposed to nutritional scarcities firstly to even identify whether these or neighboring variants show the same associations with height and BMI, with the challenge of very different genetic architecture, in terms of linkage disequilibrium patterns, compared to white European populations in whom the vast majority of GWA studies are set. Secondly, in studies of genetic selection, compared to the lactase gene variants that confer lactase persistence and the ability to digest milk beyond mid-childhood, the identified SNPs for continuous childhood growth traits have far more modest effect sizes and selection may be more difficult to prove. Future putative infrequent (non-rare) variants in these genes with greater effects on these traits may allow future test of this hypothesis.

11. Conclusions

While shorter-term adaptations in metabolic, endocrine, and epigenetic processes may occur in response to challenging environments, substantial proportions of these highly heterogenous traits are governed by genetic factors. Genetic adaptations for childhood growth, weight gain and maturation timing could have clinical significance not just for prediction and understanding of normal and abnormal variations in these childhood traits, but may also help to explain the early life links to adult disease risk. An evolutionary developmental approach to genetic epidemiology may help us to understand the wide variations in childhood growth patterns and maturation timing, and aid the prediction of the long-term consequences of modifying these traits by our endocrine treatments, public health policies, or changing nutritional environments.

D. NOTE BY MOSHE SZYF ON THE DNA METHYLATION PATTERN AS A MOLECULAR LINK BETWEEN EARLY CHILDHOOD AND ADULT HEALTH

Although epidemiological data provides evidence that there is an interaction between genetics (nature) and the social and physical environments (nurture) in human development, the main open question remains the mechanism. A coating of methyl groups covalently modifies the DNA molecule. The pattern of distribution of methyl groups in DNA is different from cell type to cell type and confers cell-specific identity on DNA during cellular differentiation and organogenesis. This is an innate and highly programmed process. However, recent data suggest that DNA methylation is not only involved in cellular differentiation but that it is also involved in the modulation of genome function in response to signals from the physical biological and social environments. We propose that modulation of DNA methylation in response to environmental cues early in life serves as a mechanism of lifelong genome adaptation that molecularly embeds the early experiences of a child ("nurture") in the genome ("nature"). Adaptations of the genome through alterations in DNA methylation are proposed to occur at different timescales from "evolutionary" to "physiological." Critical times in life when genome adaptation is speculated to occur are early life puberty and aging. There are very scant data supporting this hypothesis to date, and it needs to be tested experimentally. However, there is an emerging line of data for DNA methylation changes in response to the early life environment.

1. Introduction

One of the long-standing questions that have bewildered humans is whether "nature" or "nurture" defines our phenotype and lifelong trajectories. Modern biology has discovered the chemical molecule that embodies nature—the DNA that constitutes our genome. There has been a genocentric focus in our approach to understanding interindividual differences in human and animal behavior and physiology. The underlying hypothesis has been that differences in gene sequence between individuals result in alteration in genome function, and this is the exclusive explanation for phenotypic differences between individuals. However, a large body of literature has suggested that experiences, especially early in life, that constitute "nurture" are important in the development of both behavior and physical properties, thereby raising the question of how do these interact with the "nature" of our genes and our genome. These interactions were termed "gene by environment interactions" and were quantitatively recognized in epidemiological studies but the mechanisms of such possible interactions remained ephemeral. A classic example of gene–environment interaction is the study by Caspi et al., who have shown that a functional polymorphism in the promoter region of the serotonin transporter (5-HT T) gene moderated the influence of stressful life events on depression (Caspi, Sugden et al. 2003).

I will propose here that a chemical modification of DNA, methyl moieties distributed in a cell-type-specific manner on the DNA, is a candidate to serve as a registry of environmental exposures in the DNA molecule itself and to modulate genome function as an adaptation to these exposures. Nurture is embedded in the chemistry of DNA in the form of the pattern of distribution of methyl groups. The DNA molecule bears two layers of information in its chemical structure: the ancestral genetic information encoded in the sequence of bases in DNA and the pattern of distribution of methyl cytosines (Razin and Riggs 1980; Razin and Szyf 1984) and hydroxymethyl cytosines (Kriaucionis and Heintz 2009). The methyl groups are added to the DNA molecule after replication by machinery that is distinct from the machinery that replicates the DNA sequence that creates a potential for differential information within the hard chemistry of the DNA molecule itself (Razin and Szyf 1984). The DNA molecule has two identities: the ancestral identity encoded in the sequence and the cell-specific identity encoded in the pattern of DNA methylation. Cell-type-specific patterns of DNA methylation provide an answer to the question of how can one genome express multiple phenotypes, as is the case in multicellular organisms such as humans. It was well established that DNA methylation patterns were formed during embryogenesis by innate organized developmental programs of methylation and demethylation, creating a pattern of methylation that was cell specific (Razin and Riggs 1980; Razin and Szyf 1984). This was confirmed with modern genome-wide DNA methylation mapping methods (Lister, Pelizzola et al. 2009). Since cellular differentiation was believed to be terminal, it was also believed that DNA methylation patterns once formed remained fixed. However, recent data support the idea that DNA methylation is dynamic after birth (Szyf 2009). It will be proposed here that DNA methylation performs an additional function to its role in establishing cellular identity; DNA methylation changes in response to experience confer an "experience-based" identity to genomes. Both proposed roles of DNA methylation provide a mechanism for genome function diversification; a similar genome expresses multiple phenotypes either in different cell types through cellular differentiation or in response to different environmental exposures.

2. DNA Methylation Patterns and Their Roles in Cellular Differentiation and Gene Expression

Vertebrate DNA is covalently modified by enzymatic addition of a methyl group to cytosines in DNA (Fig. 11.D.1). The distribution of the methyl groups in the DNA varies across different cell types in an organism, creating a cell-type-specific pattern of DNA methylation (Fig. 11.D.2) (Razin and Riggs 1980; Razin and Szyf 1984). In vertebrates, the cytosine-guanine (CG) dinucleotide sequence is a principal target of DNA methylation (Gruenbaum, Stein et al. 1981) since it is preferentially recognized by vertebrate DNA methyltransferases (Gruenbaum, Cedar et al. 1982). CG is the only dinucleotide sequence that is automatically heritable from the template strand since the cytosine is in a palindrome; that is, across a methylated CG in the parental strand there would be a CG in the daughter strand (Razin and Riggs 1980). While other contexts of C such as CC could not be copied since across a CC there would be a GG in the parental strand. Heritability of the DNA methylation mark is critical for its role as an epigenetic signal that bears the specific cellular identity across cellular division. However, recent studies suggest that DNA methyation could occur in non-CG sequences at least in stem cells, but it is unclear how prevalent non-CG methylation is in postmitotic tissues (Lister, Pelizzola et al. 2009; Fuso, Ferraguti et al. 2010). In any case, the presence of non-CG methylated sites in our genomes suggests that mechanisms for the inheritance of DNA methylation exist that are not defined automatically by the DNA methylation of the parental template strand of DNA. Non-CG methylation would be perfectly in place for postmitotic cells such as neurons and could serve as a lifelong mark since there is no need to copy the mark.

DNA methylation diversifies genome function because it is generated and maintained by enzymatic machinery that is independent of the DNA replication machinery (Fig. 11.D.1). Watson and Crick rules define how DNA methylation is copied to

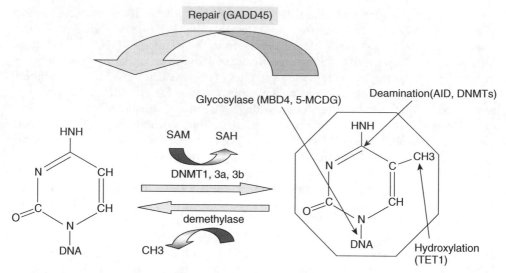

Fig. 11.D.1. DNA methylation reactions. DNA methyltransferases transfer methyl groups (CH$_3$) from the methyl donor S-adenosylmethionine (SAM) to cytosines in DNA. Several demethylation reactions were suggested. Direct demethylation by a demethylase enzyme could release a methyl moiety (CH$_3$) in the form of either methanol or formaldehyde. Alternatively, the methyl cytosine ring could be modified by either deamination catalyzed, for example, by activation-induced cytidine deaminase (AID) or by the DNA methyltransferases (DNMT), which were shown to catalyze deamination of 5-methylcytidine in the absence of SAM or hydroxylation of the methyl moiety catalyzed by TET1. The modified base is then excised and repaired. Alternatively, the bond between the sugar and the base is cleaved (by glycosylases such as MBD4 or 5-methylcytosine glycosylase 5-MCDG) followed by repair. Repair proteins shown to be associated with demethylation were GADD45(a and b).

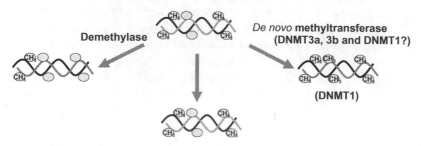

Fig. 11.D.2. Sculpting of DNA methylation patterns during cellular differentiation. Two kinds of DNA methylation reactions are shown: *de novo* methyltransferases (DNMT3a and DNMT3b) add new methyl groups to cytosines in DNA, maintenance DNA methyltransferase (DNMT1) copies the DNA methylation pattern from the template strand following DNA replication. DNA could be actively demethylated by demethylases that remove the methyl groups from DNA (see Fig. 11.D.1) resulting in different patterns of DNA methylation. Upon completion of cellular differentiation, each cell type in the body has its distinct pattern of methylation that reflects in a unique program of gene function.

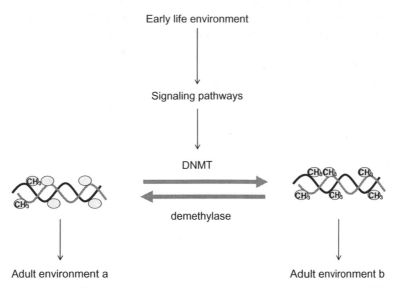

Fig. 11.D.3. The early life environment modulates the DNA methylation equilibrium. The DNA methylation equilibrium is laid down during embryogenesis by innate developmental programs. A balance of DNA methylation and demethylation activities dynamically maintains this pattern and is attuned to signals from the early environment that can modulate the pattern through activation of signaling pathways that facilitate either increased demethylation or increased methylation. The DNA methylation balance set by early life environment programs the genome to adapt to anticipated lifelong environments.

ensure the faithful and almost fixed inheritance of genetic information during cell division to all daughter cells. DNA methylation is catalyzed by an equilibrium of DNA methylating and demethylating enzymes (Fig. 11.D.1). This equilibrium could be modulated during cellular differentiation to generate cell-specific patterns of methylation and could be responsive to signals transmitted from the environment to generate an experience-responsive pattern of methylation (Fig. 11.D.3).

The DNA methylation reaction is catalyzed by DNA methyltransferases (DNMTs) (Razin and Cedar 1977). Methylation of DNA occurs immediately after replication by a transfer of a methyl moiety from the donor S-adenosyl-L-methionine (SAM, AdoMet) in a reaction catalyzed by DNMTs (Turnbull and Adams 1976) (Fig. 11.D.1). Three distinct phylogenic DNMTs were identified in mammals. DNMT1 shows a preference for hemimethylated DNA *in vitro*, which is consistent with its role as a maintenance DNMT, whereas DNMT3A and DNMT3B methylate unmethylated and methylated DNA at an equal rate, which is consistent with a *de novo* DNMT role (Okano, Xie et al. 1998) (Fig. 11.D.1). As expected, DNMT1, which is a maintenance DNA methyltransferase that copies methyl groups from the parental to daughter strand, is CG specific since this is the only C context that could be replicated. DNMT3a and DNMT3b have lax sequence specificity. Two additional DNMT homologs were found: DNMT2, whose substrate and DNA methylation activity is unclear (Vilain, Apiou et al. 1998) but was shown to methylate tRNA (Goll, Kirpekar et al. 2006; Rai, Chidester et al. 2007), and DNMT3L, which is essential for the establishment of maternal genomic imprints but lacks key methyltransferase motifs, and is possibly a regulator of methylation rather than an enzyme that methylates DNA (Bourc'his, Xu et al. 2001; Bourc'his and Bestor 2004).

For changes in DNA methylation to occur in differentiated cells, it is important that the process of maintenance of DNA methylation is not exclusively automatic. It is becoming clear now that DNMTs are targeted to specific sequences in the genome and that the targeting factors are required not only for generating the patterns of methylation but also for maintaining the pattern of DNA methylation. For example, UHRF1 (ubiquitin-like, containing PHD and RING finger domains 1), also known as NP95 in mice and ICBP90 in humans, is required for targeting DNMT1 to newly replicating hemimethylated DNA (Bostick, Kim et al. 2007). DNMTs are found in complexes with other proteins that include other chromatin modifying proteins such as histone deacetylases (HDACs) HDAC1 and HDAC2 (Fuks, Burgers, Brehm et al. 2000; Fuks, Burgers, Godin et al. 2001). The discovery that DNMT1 and other DNMTs are targeted to specific sites by chromatin modifying enzymes suggests that DNA methylation is not exclusively automatic and provides a mechanism for a targeted change in DNA methylation in response to activation of signaling pathways. For example, it was suggested that the transcription factor NGFIA targets histone acetylation and demethylation activities to the glucocorticoid receptor gene in response to maternal care (Weaver, D'Alessio et al. 2007). Transcription factors that target DNA and chromatin modifying enzymes to genes might play an important role in experience- triggered DNA methylation changes.

The most controversial issue in the DNA methylation field is the question of whether the DNA methylation reaction is reversible (Ramchandani, Bhattacharya et al. 1999). For DNA methylation to act as a responsive signal after birth, it needs to be reversible in postmitotic cells like other biological signals. Experience-based DNA methylation responses must change in either direction and since neurons are mostly postmitotic, the loss of DNA methylation should occur actively in the absence of DNA methylation. However, it was originally believed that demethylation could occur only through a "passive" process of replication in the absence of DNMTs. Several authors have provided evidence for replication-independent demethylation (Wilks, Seldran et al. 1984; Szyf, Theberge et al. 1995; Oswald, Engemann et al. 2000; Lucarelli, Fuso et al. 2001; Bruniquel and Schwartz 2003), and it has been shown that brain extracts are capable of demethylating "naked" DNA substrate *in vitro* (Mastronardi, Noor et al. 2007; Dong, Nelson et al. 2008; Fuso, Nicolia et al. 2011). The strongest evidence for dynamic methylation–demethylation comes from several studies showing active demethylation in postmitotic neurons (Weaver, Cervoni et al. 2004; Levenson, Roth et al. 2006; Miller and Sweatt 2007; Feng, Zhou et al. 2010). Conditional knock out of DNMT1 in postmitotic neurons results in DNA demethylation, suggesting the presence of demethylation activity in nondividing neurons (Feng and Fan 2009). Two kinds of mechanisms could operate in nondividing cells to remove methyl groups from DNA; removal of the methylated base followed by a repair process in the absence of DNA methyltransferase activity and a true direct removal of the methyl group by enzymatic demethylation. The most commonly accepted mechanism is repair based since it does not invoke a hitherto unprecedented enzymatic activity. Several repair-based mechanisms were proposed for the replacement of methylated DNA with unmethylated DNA. First, the methylated cytosine could be removed by a glycosylase activity. The abasic site that was created is then repaired and replaced with an unmethylated cytosine (Razin, Szyf et al. 1986; Jost 1993). Second, DNMTs were proposed to deaminate the methyl cytosine to thymidine creating a C/T mismatch, which is then corrected by a mismatch-repair mechanism (Kangaspeska, Stride et al. 2008). DNMTs were previously shown to deaminate 5-methylcytosines (Shen, Rideout et al. 1992; Zingg, Shen et al. 1998) under conditions of low SAM. Third, growth arrest and DNA-damage-inducible, alpha (GADD45A), a DNA repair protein, was proposed to participate in catalysis of active DNA demethylation by an unknown DNA repair-based mechanism (Barreto, Schafer et al. 2007). However, this was disputed (Jin, Guo et al. 2008). Other studies have suggested involvement of GADD45B in demethylation in the brain (Ma, Jang et al. 2009). Fourth, a complex sequence of coupled enzymatic reactions of deamination and mismatch repair were shown to be involved in demethylation in zebrafish: activation-induced cytidine deaminase [(AID), which converts 5-meC to thymine], a G:T mismatch-specific thymine glycosylase methyl-CpG binding domain protein 4 (MBD4) and repair promoted by GADD45A (Rai, Huggins et al. 2008). AID has been implicated in the global demethylation in mouse primordial germ cells as well (Popp, Dean et al. 2010) (Fig. 11.D.1). The main challenge in accepting a repair-based mechanism as a lifelong physiological process is that it invokes constant damage to the integrity of DNA; constant breaking and fixing of the DNA seems to be an extremely dangerous way to maintain the DNA methylation equilibrium in both the developing embryo and postmitotic neurons.

In contrast to these repair-based mechanisms, we have previously proposed that demethylation is truly a reversible reaction that involves removal of the methyl moiety rather than breaking the DNA and fixing it with an unmethylated cytosine (Ramchandani, Bhattacharya et al. 1999) (Fig. 11.D.1). We proposed that the methylated DNA binding protein 2 (MBD2) was a bona fide demethylase that removed methyl groups from DNA and truly reversed the DNA methylation reaction. This is to date the only described bona fide demethylase. MBD2 has been implicated in the activation of both methylated and unmethylated genes (Fujita, Fujii et al. 2003; Angrisano, Lembo et al. 2006). Several groups (Ng, Zhang et al. 1999; Wade, Gegonne et al. 1999) have contested the demethylase and transcriptional activating properties of MBD2. Studies by Detich et al. have demonstrated MBD2 demethylase activity *in vitro* (Detich, Theberge et al. 2002). Hamm et al. have proposed an oxidative mechanism of 5-cytosine DNA demethylation by MBD2 (Hamm, Just et al. 2008). According to this mechanism, oxidation of the methyl moiety generates 5-hydroxymethylcytosine by oxidation, which is followed by release of the methyl residue in formaldehyde.

Interestingly, 5-hydroxymethylcytosine was recently discovered in mammalian DNA (Pelizzola, Koga et al. 2008). A recent study has shown that TET1, an enzyme that converts methylcytosine to hydroxymethylcytosine, is required for maintaining the demethylated state of NANOG in mouse ES cells supporting a possible role for TET1 and 5-hydroxymethylcysoine as an intermediary in the demethylation reaction (Ito, D'Alessio et al. 2010). However, more recent data suggest that 5-hyroxymethylcytosine is stable in DNA and is especially abundant in the brain, possibly playing a role in the state of activity of genes (Jin, Wu et al. 2011). Future analysis of DNA methylation must take into account the two states of DNA modification 5-hydroxymethylcytosine and methyl cytosine. In summary, although there is no agreement as of yet on the mechanism of DNA demethylation, the presence of active demethylation in somatic cells is widely acknowledged.

What is the information that is conveyed by the DNA methylation pattern? If DNA methylation is playing a role in diversifying the number of phenotypes expressed by one genome, then it should affect stable programs of differential gene expression. Razin and Riggs proposed three decades ago that DNA methylation silences gene expression and therefore patterns of DNA methylation could be used to differentially silence parts of the repertoire of genes in the genome (Razin and Szyf 1984; Razin and Riggs 1980). There is an overall inverse correlation between DNA methylation in several regulatory regions of genes and gene expression, which was discovered in the early 1980s and was confirmed by whole genome approaches (Rauch, Wu et al. 2009). However, exceptions to this rule exist and unmethylated promoters of genes that are silenced are frequent in the genome. It is possible that methylation-dependent regulatory regions that were not examined in published studies control these genes. Alternatively, it is important to note that DNA methylation and gene expression studies describe different levels of gene expression regulation. DNA methylation patterns program gene expression potential and condition genes to respond to appropriate environmental and physiological signals, whereas expression studies reveal a transient state of expression.

Several lines of study have provided an understanding of the mechanisms that are involved in gene silencing by DNA methylation. Highly methylated DNA regions are packaged in the nucleus in inactive chromatin configuration (Razin and Cedar 1977). These data point to a relationship between two critical elements of the epigenome: chromatin modification and DNA methylation. Several studies support a bilateral relationship between DNA methylation and chromatin structure; DNA methylation directs the formation of inactive chromatin while inactive chromatin targets DNA methylation. The converse is also true: loss of methylation targets activating histone modification marks, and active histone methylation marks such as histone acetylation facilitate DNA demethylation (D'Alessio and Szyf 2006). At least two mechanisms were demonstrated for inhibition of gene activity by DNA methylation. A methyl group positioned in a recognition element for a transcriptional factor can block binding of the transcription factor to the promoter (Comb and Goodman 1990; Inamdar, Ehrlich et al. 1991). Alternatively, methylated DNA attracts methylated DNA binding proteins (MBD) such as the Rett syndrome protein methyl CpG binding protein 2 (MeCP2), which in turn recruits histone modification enzymes such as HDACs to the gene, precipitating an inactive gene-silencing chromatin configuration (Nan, Campoy et al. 1997). Although the main focus of most studies has been on promoter DNA methylation, recent whole genome methylation analyses revealed an interesting positive relationship between DNA methylation in the bodies of genes and gene expression (Hellman and Chess 2007; Lister, Pelizzola et al. 2009; Rauch, Wu et al. 2009). A large fraction of the methylation in DNA occurs in intergenic regions and in repetitive sequences, and this methylation might play yet unknown roles. Methylation of retroviral elements in the genome was proposed to silence ectopic expression from these parasitic elements that could disrupt normal genome function (Yoder, Walsh et al. 1997).

An open question is the role of the newly discovered modification hydroxymethylcytosine (Kriaucionis and Heintz 2009). Although originally it was proposed to be an intermediary in DNA demethylation it appears that this is a stable modification of DNA that tends to be concentrated in active gene promoters (Jin, Wu et al. 2011). TET enzymes catalyze hydroxymethylation of cytosine only once they are methylated by DNMTs (Ito, D'Alessio et al. 2010). Hydroxymethylation might serve as a mechanism for activating a gene without losing the DNA methylation mark and might serve an important role in the long-term preservation of methyl marks. Others speculate that it serves as a mechanism for activation of genes that were aberrantly silenced by DNA methyation as a guardian against unscheduled silencing of gene expression. Future studies in DNA methylation and the role that it plays in response to the environment will have to consider the role of 5-hydroxymethylcytosine.

3. DNA Methylation as a Genome Adaptation Mechanism

It is well established that DNA methylation plays a critical role in providing different identities to the same genome during cellular differentiation. But this process of differentiation of the DNA methylation pattern is innately programmed during development. Highly organized processes drive it during embryogenesis. The critical question is whether DNA methylation also participates in phenotypic plasticity that is observed during postnatal development, a process that shares with embryogenesis the need to express multiple phenotypes from similar genomes but is different from differentiation, which responds to innate cues it needs to respond to changing external environmental cues including the physical, biological and social environments. Is there a mechanism that responds to cues from the environment to adapt the genome to changing environments? DNA methylation is an excellent candidate to participate in these processes since it is generated and maintained by enzymes that could be potentially modulated. Although genetic alteration is a mechanism for adaptation through selection, such a mechanism is unrealistic within a span of a lifetime or few generations. The genome maintains a rigid source of functional information in a highly dynamic environment creating a serious challenge for adaptation to changing environments. We propose that DNA methylation serves as an interface between the dynamic environment and the fixed genome.

We suggest that DNA methylation can participate in genome adaptations at multiple timescales: from an evolutionary timescale to a proximal physiological timescale. Particularly critical are early life events when information from the social, bioenvironment, and physical environments serve to modulate the phenotype and lifelong trajectories of health. It is proposed that these DNA methylation adaptations early in life are system wide and that they involve multiple gene circuitries (Fig. 11.D.3). It is proposed that a DNA methylation genome-wide adaptation can turn "maladaptive" under specific contexts later in life and result in mental health pathology. Similar critical windows are proposed to be transitions in life such as weaning, puberty, and aging. At these points the DNA methylation machinery might be responsive to signals from the environment for long-term programming of states of methylation.

It is interesting to note that there is evolutionary evidence for involvement of DNA methylation in sociality especially in the caste structure in the honeybee (Kucharski, Maleszka et al. 2008; Maleszka 2008; Lyko, Foret et al. 2010). It is possible that DNA methylation evolved to translate social information to genomic programs of social behavior.

Epidemiological data point to the importance of early life experience in setting lifelong health and mental health trajectories in humans (Power, Jefferis et al. 2006). It is therefore hypothesized that early life is an especially sensitive time for DNA methylation adaptation in the brain and system wide. It is predicted that an "adaptation" mechanism leading to pathology will involve several gene circuits and a system-wide adaptation of the DNA methylation pattern to an anticipated environment.

An interesting question is whether DNA methylation differences between individuals are stochastic or are an organized adaptive response or a combination of both. The existence of stochastic differences in DNA methylation in the germ line and existence of such differences in DNA methylation even between identical twins were recently demonstrated (Kaminsky, Wang et al. 2006; Petronis 2006; Mill, Tang et al. 2008; Kaminsky, Tang et al. 2009; Flanagan, Popendikyte et al. 2006). Stochastic responses might result from environmental encounters that affect DNA methylation enzymes. Increased or decreased methyl content in the maternal diet would affect the phenotype of the offspring (Waterland and Jirtle 2003; Dolinoy, Weidman et al. 2006; Dolinoy, Das et al. 2007; Dolinoy, Huang et al. 2007; Dolinoy and Jirtle 2008). One can envision that such diets would have a global effect on DNA methylation enzymes resulting in stochastic loss of methyl groups. Alternatively, reduction or increase in methyl content might be sensed by signaling pathways in the cell that will evoke an organized response that will result in a distinct response of the methylome. There are important implications on diagnostics prevention and therapy if the DNA methylation response to the environments is indeed an adaptive response. Future studies will need to address this question.

4. Epigenetic Programming by the Early Life Social Environment

Perhaps one of the most remarkable examples of how environments affect development after birth is the impact of the early life social environment on health trajectories later in life (Power, Hertzman et al. 1997; Hertzman, Power et al. 2001; Power, Li et al. 2006). If the social environment affects DNA methylation, its impact cannot result from stochastic inhibition of DNA methylation/demethylation enzymes; that could be an explanation for the impact of toxins or food ingredients but not social exposures. Social environments must evoke signaling pathways in the brain and the body that are associated with organized responses. There are several models that measure the impact of early life social environment on behavior and other health phenotypes later in life. Animal models could be used to test whether the impact of early life social environment on the phenotype is mediated by "genetic" or "epigenetic" mechanisms. Maternal behavior plays a cardinal role in the behavioral development of mammals. Models of maternal deprivation in primates and rodents, and natural variation in maternal care in rodents, were used to demonstrate the profound impact of maternal care and "nurture" on a panel of phenotypes in the offspring that last into adulthood (Ruppenthal, Arling et al. 1976; Suomi, Collins et al. 1976).

Hippocampal glucocorticoid receptor (GR) controls the negative feedback of the HPA axis by glucocorticoids. In the rat, the adult offspring of mothers that exhibit increased levels of pup licking/grooming (i.e., high-LG mothers) over the first week of life show increased hippocampal (GR) expression, enhanced glucocorticoid feedback sensitivity, decreased hypothalamic corticotrophin releasing factor (CRF) expression and more modest hypothalamic–pituitary–adrenal (HPA) stress responses compared to animals reared by low-LG mothers (Liu, Diorio et al. 1997; Francis, Diorio et al. 1999). The *GR/NR3C1* gene encoding the glucocorticoid receptor (GR exon 17 promoter) exhibits differences in DNA methylation and histone acetylation in the hippocampus of the offspring of high- and low-LG mothers. Differences in epigenetic programming in response to differences in maternal LG emerged early in life and remained stable into adulthood, illustrating how epigenetic programming early in life could set up lifelong behavioral trajectories (Weaver, Cervoni et al. 2004).

The basic concepts of this study were repeated more recently in several other models of early life social adversity. Exposure of infant rats to stressed caretakers that displayed abusive behavior produced persisting changes in methylation of BDNF gene promoter in the adult prefrontal cortex (Roth, Lubin et al. 2009). Early life stress (ELS) in mice caused sustained DNA hypomethylation of an important regulatory region of the arginine vasopressin (AVP) gene (Murgatroyd, Patchev et al. 2009).

An extremely important question is whether the results in rodents could be translated to humans? The state of methylation of rRNA gene promoters and GR were examined in a cohort of suicide victims in Quebec who were abused as children, and their control group. Ribosomal RNA (rRNA) forms the skeleton of the ribosome, the protein synthesis machinery. Protein synthesis is essential for building new memories and creating new synapses in the brain. Our genome contains around 400 copies of the genes encoding rRNA. One possible way to control the protein synthesis capacity of a cell is through changing the fraction of active rRNA alleles in a cell (Brown and Szyf 2007). We have previously shown that the fraction of rRNA genes that is active and is associated with the RNA Pol1 transcription machinery is unmethylated, while the fraction that is inactive is methylated (Brown and Szyf 2007). Our results showed that the suicide victims who experienced childhood abuse had higher overall methylation in their rRNA genes and expressed less rRNA. This difference in methylation was region specific: It was present in the hippocampus and was not observed in the cerebellum. Moreover, although significant methylation differences were observed between the controls and the suicide victims, no sequence differences were observed. The fact that the difference in methylation was brain-region specific and that no sequence differences were observed further strengthens the conclusion that this difference in methylation was driven by environmental rather than genetic variation (McGowan, Sasaki, Huang et al. 2008). These data point to the possibility that the effects of early life adversity might not be limited to the usual suspects of highly brain specific genes but that ubiquitously expressed genes could be involved as well. Modulation of expression of ubiquitous genes might be important in modulating brain function.

Individuals with treatment-resistant forms of major depression show decreased GR expression and increased HPA activity. Site-specific differences in DNA methylation in the GR exon 1f promoter and its expression were detected between suicide completers who had reported social adversity early in life and suicide completers who did not experience social adversity early in life (McGowan, Sasaki,

D'Alessio et al. 2009). Differences in DNA methylation of the GR promoter were observed also in peripheral blood cells; the GR promoter was more methylated in lymphocytes in newborns exposed prenatally to maternal depression than control newborns (Oberlander, Weinberg et al. 2008). This lends support to the hypothesis that DNA methylation differences in response to social adversity are system wide and are not limited to brain specific regions.

Epigenetic modulation of other candidate genes was implicated in suicide; the gamma-aminobutyric acid A receptor alpha 1 subunit (GABRA1) promoter (Linthorst, Flachskamm et al. 1995) within the frontopolar cortex (Poulter, Du et al. 2008) and tropomyosin-related kinase B (TRKB) in the frontal cortex of suicide completers (Ernst, Deleva et al. 2009). It is unknown yet whether these changes in DNA are also associated with early life adversity.

5. Genome and System-Wide Impact of Early Life Adversity

The first studies summarized above focused on a candidate gene approach. However, the large number of phenotypes that are associated with early life adversity both in animals and humans suggest that the impact of early adversity on the DNA methylation pattern will be broad. Moreover, it is clear that genes do not act independently but through functional gene circuitries. We therefore reasoned that adaptation of the DNA methylation pattern to early life adversity will be broad and that it will involve several systems in the body. We tested this hypothesis in several studies.

First, we examined the state of DNA methylation, histone acetylation and gene expression in a 7 million base pair region of chromosome 18 containing the GR gene in the hippocampus of adult rats and showed that natural variations in maternal care in the rat are associated with coordinate changes in DNA methylation, chromatin, and gene expression spanning over a hundred kilobase pairs. Interestingly, a chromosomal region containing a cluster of the protocadherin α, -β, and -γ (Pcdh) gene families implicated in synaptogenesis show the highest differential response to maternal care. The entire cluster reveals epigenetic and transcriptional changes in response to maternal care. These studies suggest that the DNA methylation response to early life maternal care is coordinated in clusters that cover broad areas in the genome and that the epigenetic response to early life maternal care involves not only single candidate gene promoters but includes transcriptional and intragenic sequences, as well as those residing distantly from transcription start sites and regions containing noncoding RNAs (McGowan, Suderman et al. 2011).

Second, we showed that a similar pattern of response to childhood abuse is associated with DNA methylation differences throughout the genomic region spanning the six and a half million base-pair region centered at the *NR3C1* gene in the hippocampus of adult humans. The DNA methylation differences associated with child abuse bear a striking resemblance to DNA methylation differences between adult offspring of high- and low-maternal-care rats. This provides evidence for an analogous cross-species epigenetic and transcriptional response to early life environment (Suderman et al., submitted 2011).

Third, we tested whether the response to early life adversity is system wide and includes T cells as well as the brain by examining in parallel the impact of differential maternal rearing in a rhesus model of maternal deprivation. We examined the impact of depriving maternal care on DNA methylation in the prefrontal cortex and T cells. Our results show that similar to the rat and human, the changes associated with differences in rearing are widespread in the genome and that they are not limited to the brain and occur in T cells as well. Although the vast majority of DNA methylation changes that associate with rearing are different in T cells and prefrontal cortex, some similarities were detected. These data are consistent with the hypothesis that the response to early life adversity is genome wide and system wide. Since changes in DNA methylation associated with maternal rearing were detected in T cells, it might be possible to perform population DNA methylation studies examining either whole blood or T cells. We have initiated a study of the impact of socioeconomic positioning on DNA methylation that examined blood DNA from the British birth cohort of 1958. This study detected a signature of DNA methylation that is associated with early life adversity (Borghol et. al., unpublished).

6. Prospective and Summary

DNA methylation changes play a role in the highly programmed innate process of cellular differentiation by conferring different identities to the same genetic sequence. The process of cellular differentiation teaches us that it is possible for identical genomes to have different chemical identities defined by covalent chemical modifications. The phenomenon of phenotypic plasticity suggests that there must be mechanisms that allow identical genomes to express different phenotypes in response to environmental challenges. The critical question is mechanism. This chapter suggests that the same mechanism that provides genomes with different identities during cellular differentiation is involved in conferring differential identities to individuals in response to environmental cues. This hypothesis implies that there are different processes that could alter DNA methylation in addition to the highly programmed process of cellular differentiation and that the DNA methylation is far more plastic than originally thought. Involvement of DNA methylation in terminal differentiation requires that the DNA methylation pattern will be fixed in mature cell types and consistent among individuals. A DNA methylation that is responsive to the environment will imply a more dynamic situation. We suggest that these changes in DNA methylation in response to the environment constitute an adaptive response of the genome rather than stochastic drifts in DNA methylation. We also propose that these could turn maladaptive when there is a misfit between the DNA methylation pattern and the environment, resulting in human disease.

Our challenge is to understand the mechanisms that direct such genome wide and system-wide DNA methylation adjustments and to delineate how these changes in DNA methylation could lead to changes in genome function and physiology. An additional challenge is to delineate the time points in life when DNA methylation adaptation occurs. Although there are data suggesting that the DNA methylation pattern responds to adversity early in life, it stands to reason that there are other critical time points in life when DNA methylation changes could have a long-term impact on programming of physiological function. DNA methylation signatures could serve as predictors of human disease and serve as critical tools in identifying subjects at risk as well as in guiding prevention and treatment strategies. If indeed DNA methylation adaptations are system wide, then DNA methylation changes in blood might serve as markers of behavioral and physiological disorders. DNA methylation signatures are attractive candidates for novel molecular diagnostics and are bound to change the face of this field.

DNA methylation patterns that are different from genetic mutations are reversible, and the possibility that DNA methylation changes are driving human disease and health problems has an optimistic message; it might be possible to reverse maladaptive DNA methylation marks either with pharmacological or behavioral interventions.

12

LIFE HISTORY THEORY IN UNDERSTANDING GROWTH DISORDERS

Each human subject has a slightly different genetic composition, evolutionary history, and life history. The following examples demonstrate how the understanding of life-history theory helps in understanding growth mechanisms in a given child.

The life-history approach of infancy–childhood transition has previously been shown to be helpful in analyzing the growth failure of girls with Turner's syndrome (Karlberg, Albertsson-Wikland et al. 1991; Even, Cohen et al. 2000). Using growth data from three well-defined syndromes of growth retardation, I present here life-history stage transition as it is utilized to analyze child growth in health and disease, and its contribution to understanding their mechanisms: Down syndrome, Noonan's syndrome, and Silver-Russell syndrome.

Down syndrome is caused by trisomy of all or a critical portion of chromosome 21. It is characterized mostly by mental retardation, hypotonia, short stature, and characteristic faces, but there are often also congenital heart malformations and malformations of the gastrointestinal tract.

Noonan syndrome is distinguished by characteristic faces, short stature, short neck with webbing or redundancy of skin, cardiac anomalies, motor delay, and a bleeding diathesis.

Silver-Russell syndrome was described independently by Silver et al. in 1953 and by Russell in 1954 as a syndrome of low birth weight, severe growth retardation and short stature, characteristic facial features and congenital hemihypertrophy. Most Silver-Russell children have severe feeding problems, sweating, and pallor in the early weeks of life, and about half are considered for special education.

The mechanisms of short stature in all three syndromes are poorly understood, and the following is an attempt to use life-history theory to understand them. The

Evo-Devo of Child Growth: Treatise on Child Growth and Human Evolution, First Edition.
Ze'ev Hochberg.
© 2012 Wiley-Blackwell. Published 2012 by John Wiley & Sons, Inc.

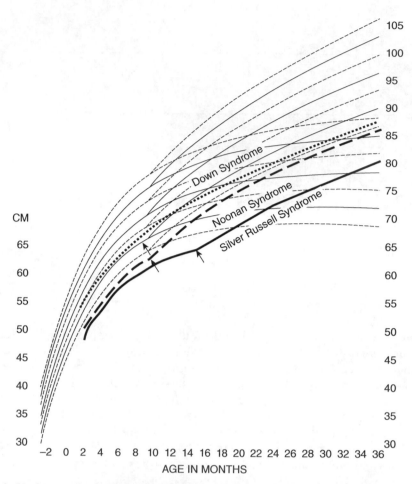

Fig. 12.1. Understanding growth failure mechanisms by analyzing the transition from infancy to childhood in girls with three common growth disorders: Down syndrome [data from Cronk et al. (1988). (dotted line)], Noonan syndrome [data from Ranke et al. (1988) (dashed line)], and Silver-Russell syndrome [data from Wollmann et al. (1995) (solid line)], as compared with the normal transition age of 6–12 months (mean ± 2 SD, gray box). The arrows indicate the normal mean age of transition for Down syndrome at 8.8 months and Noonan syndrome at 9.3 months, and the delayed mean transition age of Silver-Russell syndrome at 15.2 months.

mean growth values for girls with each of the three syndromes are plotted in Figs 12.1, 12.2, and 12. 3 against normal controls, and the growth attainment and loss in each stage are summarized in Table 12.1.

A. DOWN SYNDROME

The average female infant with Down syndrome has a mild growth failure during her initial year of life, and loses 0.5 SDS over that period from a mean –1.2 SDS at age 2 months to –1.7 SDS at the infancy–childhood transition (Cronk, Crocker

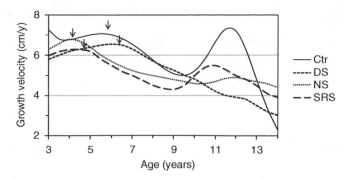

Fig. 12.2. Understanding growth failure mechanisms by analyzing life-history transitions. Linear growth (upper panel), growth velocity (middle panel), and growth acceleration (lower panel) of girls with Down syndrome (wide-dashed line), Noonan syndrome (dotted line), and Silver-Russell syndrome (narrow-dashed line), as compared with the normal (solid line). The childhood–juvenility transition ages (arrows) are 6.2, 4.1, and 4.4 years, respectively, as compared with the mean normal of 5.8 years. Adapted from Hochberg (2008).

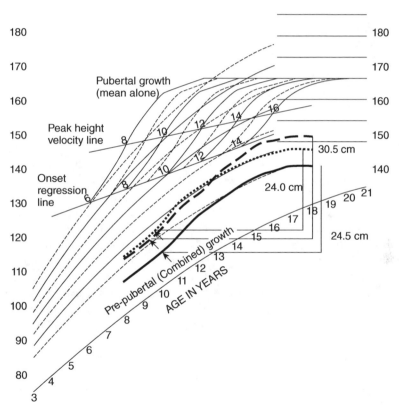

Fig. 12.3. Understanding growth failure mechanisms by analyzing life-history transitions. The juvenility–adolescent growth transition in girls with Down syndrome (dotted line) Noonan syndrome (dashed line) and Silver-Russell syndrome (solid line) as compared with the normal transition age of 8.1–12.2 years (mean ± 2 SD, gray box). The arrows indicate the normal mean age of transition for Down syndrome at 9.8 years, Noonan syndrome at 9.5 years, and Silver-Russell syndrome at 10.1 years. The total adolescent mean growth gain is shown to be 24.0, 30.5, and 24.5 cm, respectively, as compared with the normal 19–33 cm (mean ± 2SD).

181

TABLE 12.1. Impact of Life-History Stages on Growth in Girls With Down Syndrome, Noonan Syndrome, and Silver-Russell Syndrome*

	Down Syndrome	Noonan Syndrome	Silver-Russell Syndrome
Intrauterine growth (cm)	49	45	43
Intrauterine loss (cm)	*1*	*5*	*7*
Infancy growth (cm)	17	21	21
Infancy loss (cm)	*3*	*−1*	*−1*
IC transition age (months)	8.5	9.5	15
IC transition loss (cm)	*−0.4*	*0.4*	*5.4*
Childhood growth (cm)	33	22	18
CJ transition age (y)	6.2	4.0	4.3
CJ transition loss (cm)	*−1.5*	*12.2*	*9.9*
Childhood duration (y)	5.5	3.2	3.0
Juvenility growth (cm)	22	37	33
JA transition age (y)	9.4	9.7	10.0
Juvenility loss (cm)	*7.0*	*−3.0*	*−2.0*
Adolescence growth (cm)	24.0	30.5	24.5
Adolescence loss (cm)	*4.0*	*−2.5*	*3.5*
Final height (cm)	149	147	140.5
Calculated total loss (cm)	*12.9*	*13.2*	*22.8*

*Data are compared to those of an average girl growing along the 50th percentile.

et al. 1988) (Fig. 12.1). She then transitions into the childhood growth stage at a normal mean age of 8.9 months. It is during childhood growth that she loses as much as 0.8 SDS in height (Fig. 12.2). Transition of the average Down syndrome girls into juvenility is only 4 months beyond the normal control at age 6.2 and 5.8 years, respectively, and their juvenile growth deceleration are comparable. After a juvenility period of 3.6 years (4.2 years in control), the average Down syndrome girl starts her juvenility–adolescence growth transition at a normal mean age of 9.8 years (10.2 years in control) as assessed from the infancy–childhood–puberty model (Fig. 12.3); however, the intensity of the pubertal growth spurt is minimal, noticeable only at the second derivative acceleration curve (Fig. 12.3); her total pubertal growth is 24 cm, as compared with 28 cm in the average control girl. Summing her calculated loss explains 12.9 cm out of her 17-cm deficit she has in comparison with normal control 50th percentile (Table 12.1).

It is quite remarkable how little is known about the mechanism of growth failure in the familiar Down syndrome. Female infants with Down syndrome lost 1.2 SDS before they were born and additional 0.5 SDS during infancy, suggesting an insufficiency of the mechanisms responsible for intrauterine growth.

The infancy growth stage and transition to childhood were normal in Down syndrome, suggesting intact infantile growth control mechanism—a normal and timely switch turn-on of the growth hormone axis. Yet, during childhood, growth decelerates, suggesting a partial defect in this axis activity; indeed growth hormone "subnormality" has been offered as a mechanism for the growth disorder of Down syndrome (Ragusa, Romano et al. 1992).

Transition of the average Down syndrome girls into juvenility is normal at age 5.5 years, suggesting a normal response of that switch mechanism—apparently a normal maturity of the adrenal cortex reticualris and androgens generation.

Transition from juvenility into adolescence growth is also normal, indicating a normal neuroendocrine switch mechanism for puberty onset. However, the intensity of the pubertal growth spurt is minimal. Pubertal growth in girls is a function of increased activity of the growth hormone axis with the biphasic effect of estrogens: enhancing growth by low estrogen levels and suppressing effect of high estrogen concentrations. Subnormal pubertal growth in Down syndrome may indicate another possible effect of subnormal growth hormone action, or an incomplete growth effect of the combined growth hormone and pubertal estrogens. It remains to be investigated how a triple dose of chromosome 21 genes leads to this package of growth disorder in Down syndrome.

B. NOONAN'S SYNDROME

After severe intrauterine growth failure, the average Noonan female infant maintains normal growth rate along the −3.5 SDS line, and transits into the childhood growth stage at a normal mean age of 9.3 months (Ranke, Heidemann et al. 1988) (Fig. 12.1). During childhood, she maintains her height at the −3.5 SDS line, and the childhood–juvenility growth transition occurs at a remarkably early age of 4.0 years, giving her a short childhood period of only 3.2 years and a long juvenility period of 5.4 years (Fig. 12.2). Short childhood costs her a loss of 12.2 cm before she reaches the juvenility–adolescence growth transition at a relatively early age of 9.4 years (Fig. 12.3). She then has a remarkably good adolescence growth of 30 cm. Summing her calculated loss explains 13.2 cm out of her 19-cm deficit she has in comparison with normal control 50th percentile (Table 12.1).

Gain-of-function mutations in the PTPN11 gene account for over half the Noonan's patients studied. This gene encodes for protein-tyrosine phosphatase, which is known to regulate the responses of eukaryotic cells to extracellular signals (Dechert, Duncan et al. 1995), and the disease is caused by its gain of function. An early juvenility, longer pubertal growth and excessive pubertal gain of this gain-of-function mutation are intriguing. A possible result of such a package would be juvenile and adolescent compensation for an intrauterine insult. Indeed, in 30% of adult Noonan's subjects, height is in the normal range between the 10th and 90th percentiles, and only half of the females and nearly 40% of males have an adult height below the third percentile (Noonan, Raaijmakers et al. 2003); the presence or severity of heart disease was not a factor, and none of the adults with normal height had been treated with growth hormone (Ranke, Heidemann et al. 1988).

C. SILVER-RUSSELL SYNDROME

The average girl with Silver-Russell syndrome (Wollmann, Kirchner et al. 1995) has a severe form of intrauterine growth retardation and much delayed infancy–

childhood transition (DICT), with a mean transition at age 15 months (Fig. 12.1). Hence, her infantile length falls from −3.0 SDS at age 6 months to −3.5 at a normal transition age of 9 months, and she starts her childhood with a length SDS of −4.2. She has a short childhood growth stage of 3 years of a quasi-normal growth rate, which causes a loss of an additional 9.9 cm, followed by an early transition to juvenility at age 4.4 years (Fig. 12.2). Her juvenility–adolescence growth transition is normal at age 10.4 years, followed by a long and low-amplitude adolescence growth. Summing her calculated loss explains 22.8 cm out of her 25.5-cm deficit she has in comparison to normal control 50th percentile (Table 12.1).

A mechanism for the delay in this heterogeneous syndrome, which probably does not result from a single genetic defect, does not emerge from this analysis, but considering the difference between the three syndromes, it is noteworthy that infants with Silver-Russell syndrome have marked asthenia and severe feeding problems (Anderson, Viskochil et al. 2002), which may contribute to the late transition age by virtue of energy-related cues. Recent understanding of the pathogenesis of Silver-Russell syndrome suggests two molecular mechanisms, both related to loss of imprinted paternal methylation at cheomosome 7p13-p11.2. The 11p15 epimutation of the H19 promoter described in individuals with Silver-Russell syndrome is the exact mirror image of one of the molecular defects responsible for the overgrowth Beckwith-Wiedemann syndrome (exomphalos, macroglossia, and gigantism in the neonate) (Gicquel, Rossignol et al. 2005). Interestingly, in most Beckwith-Wiedemann children, growth velocity remained above the 90th percentile up to juvenility's 4–6 years of age (Sippell, Partsch et al. 1989), offering also a mirror image of the pattern described here for Silver-Russell syndrome.

D. ADDITIONAL CASES

Utilizing this approach offers means for understanding growth pattern of individual children. Table 12.2 demonstrates three girls' histories as they emerge from their growth charts in the case of Marfan syndrome, Turner syndrome, and a small-for-gestational-age girl who showed no catch-up growth.

The girl with Marfan syndrome (tall stature, disproportionately long limbs and digits, anterior chest deformity, mild to moderate joint laxity, and a heart malformation) ended up tall as expected. The syndrome is caused by mutations in the fibrillin-1 gene, which is the major constitutive element of extracellular microfibrils, and has widespread distribution in both elastic and nonelastic connective tissue throughout the body. The patient had excessive prenatal longitudinal growth—a birth length of +1.2 SDS—and she remained on the same SDS line throughout her infancy. Her normal infancy–childhood growth transition, suggesting an intact on switch of the growth hormone axis, was followed by accelerated growth during childhood from 1.2 to 2.2 SDS, indicating that her childhood juvenility growth transition was early, at age 4.1 years, giving her a loss of 8 cm during childhood. This was followed by further acceleration from 2.2 to 2.8 SDS during juvenility, which gave her a gain of 8 cm during this life-history stage. The juvenility–adolescence growth transition was also relatively early at 8.5 years, and growth was quite normal during adolescence. Summing her calculated gain/loss explains how this child ended up at the same +1.2 SDS she was born with (Table 12.2).

TABLE 12.2. Growth Analysis of Three Female Patients With Growth Disorders

	F: Marfan S	F: Turner S	F: SGA[1]
Intrauterine growth (cm)	52.5	48	45
Intrauterine loss (cm)	*−2.5*	*2*	*5*
Infancy growth (cm)	21.5	19 (−1.2 → −2.1 SDS)	23
Infancy loss (cm)	*1.5*	*1*	*−4*
IC transition age (months)	10	10	14
IC transition loss (cm)	*0.2*	*0.2*	*4*
Childhood growth (cm)	37 (1.2 → 2.2 SDS)	33	26
CJ transition age (y)	4.1	6.0	5.0
Childhood loss (cm)	*8*	*7.0*	*9*
Childhood duration (y)	3.3	4.8	3.8
Juvenility growth (cm)	33 (2.2 → 2.8 SDS)	29	22
JA transition age (y)	8.5	14 E2 therapy	10
Juvenility loss (cm)	*−7*	*−8*	*−1*
Adolescence growth (cm)	30	10	28
Adolescence loss (cm)	*−1*	*17*	*−1*
Final height (cm)	175	139	143
Calculated total loss[2] (cm)	*1*	*17*	*7*

[1]SGA = Small-for-gestational-age.
[2]Total height losses are calculated as compared to patients born with the same length SDS.

Marfan syndrome is mostly associated with tall stature, but the case presented shows she had a final height that was appropriate for her birth length. Interestingly, this child made most of her advance in growth during juvenility, a period of otherwise decelerating growth that is associated with adiposity rebound and adrenarche (Hochberg 2008, 2009). The thinness of Marfan patients may suggest a mechanistic association between a lesser growth deceleration and a lesser adiposity rebound during Marfan's juvenility. Over half a century ago, McKusick suggested that an unleashing of the normal control on longitudinal growth as a result of the defect in the fibrous elements of the periosteum might be the mechanism for tall stature (McKusick 1956).

The girl with Turner syndrome (short stature, ovarian failure, and additional malformation due to partial or complete absence of one X chromosome) had a birth length of −1.2 SDS, and lost an additional 0.9 SDS during infancy. Normal infancy–childhood growth transition was previously reported for Turner syndrome (Hochberg, Khaesh-Goldberg et al. 2005) and is followed by decelerated growth during childhood that cost her 7 cm; loss during the childhood stage suggests a deranged growth hormone–insulin-like growth factor activity, as previously reported (Hochberg, Aviram et al. 1997) and discussed (Even, Cohen et al. 2000). Her childhood–juvenility transition was normal, and juvenility was longer than normal, as puberty had to be induced at age 14, which gave her a gain of 8 cm during this life-history stage. Her poor adolescence growth on low-dose estradiol therapy made her lose additional 17.2 cm.

The third girl, who was born small-for-gestational age (SGA) and had no catch-up growth, had a birth length of −2.9 SDS, and delayed infancy–childhood growth transition, which made her lose additional 4.5 cm. The present analysis shows that the well-documented premature adrenarche (van Weissenbruch 2007) had a costly price of 9 cm in this SGA girl. Juvenility and adolescence were on time and of normal magnitude. Summing her calculated loss explains all 7 cm of her height loss against a child who started and ended at the same birth length SDS.

13

WHEN THE PACKAGES DISINTEGRATE

In developed countries there has been a marked secular trend for increased height and a reduction in the age of pubertal onset and menarche over the past 150 years. This trend is generally understood as being an expression of improved nutrition and general health in children. The evolutionary theory of life history predicts that improved nutrition and general health will result in a reduction in juvenile mortality, increased height and reduced age of menarche, and reproductive competence.

Yet, transitions from one life-history stage to the next are linked to social behaviors and maturation in *H. sapiens* as much as it is in other mammals. This synchrony has evolved because transitions from one to the next life-history stages are energetically important, and it would be disadvantageous for social maturation to precede the growth and physical maturation, or the other way around.

Thus, the timing of the infancy to childhood transition evolved to match weaning from breast-feeding, which in modern society is a much unreliable point in time. Transition from childhood to juvenility evolved to match a child's independence for food and safety provision, which are no longer required in modern society. And puberty evolved to match the age of reproduction to the social ecology of the evolving human. Yet, whereas psychosocial maturation matches the cessation of growth and reproductive competence among contemporary hunter-gatherer societies (Walker, Gurven et al. 2006), they certainly mismatch among modern humans in industrial societies (Gluckman and Hanson 2006).

I have previously defined the package we call infancy by deciduous dentition, feeding by maternal lactation, rapid and decelerating growth, and the "mini-puberty"

Evo-Devo of Child Growth: Treatise on Child Growth and Human Evolution, First Edition.
Ze'ev Hochberg.
© 2012 Wiley-Blackwell. Published 2012 by John Wiley & Sons, Inc.

surge in sex hormones. The growth transition from infancy to childhood that is associated with the setting in of the growth hormone axis activity occurs in Sweden and Israel at 9 and 10 months, respectively (Hochberg and Albertsson-Wikland 2008), but mini-puberty proceeds in girls to age 2 years and in boys to age 6 months (Burger, Yamada et al. 1991; Chellakooty, Schmidt et al. 2003). The implications of disintegration of the infancy stage—the mismatch between weaning and infantile growth—have been discussed in Chapter 4, Section F.

The transition from childhood to juvenility is part of the strategy in the transition from a period of total dependence on the family and tribe for provision and protection into self-supply; it is assigned with a predictive adaptive response of body composition and energy metabolism. To fulfill this task, the package includes molar dentition, adiposity rebound, deceleration of growth, and the provision to the brain of the neurosteroid DHEAS—the so-called adrenarche. Juvenility is endowed with programming/predictive adaptive responses for a thrifty phenotype, metabolism, and body composition. In a stable environment, evolutionary pressures operate to select traits that match the organism to its environment. It seems probable that the timing of adrenarche, growth deceleration, and adiposity rebound would be linked to social maturation associated with preparation for the next life-history stage: adolescence. Thus, the timing of the transition to juvenility evolved to match the biological maturation to the social ecology of the evolving human, and as mentioned, the juvenility age has remained constant, for at least the past 1.9 million years. This synchrony would have been selected because the energy-consuming brain has now reached both its quasi-final size and maturation, and energy stores for future energy-consuming adolescent growth spurt. To prevent social competition, it would be disadvantageous for growth to precede the capacity for sexual maturation. This is a coherent package that mostly remains intact. However, disintegration is observed in children with precocious juvenility (Hochberg 2010), as discussed in Chapter 6, Section G. Overweight and obesity that characterize modern society results in early adiposity rebound (Cole, Freeman et al. 1995; Hochberg 2008). With weight being most probably among the major signals for biological juvenility, growth deceleration, and adrenarche, as evident from studies in the low birth weight–precocious pubarche–PCOS complex. DHEA effects on a wide variety of physiological systems, including neurological and mood modulator, immune, and somatic growth and development (Campbell 2006). It would be unprecedented in light of the stable juvenility age over millions of years. What doesn't change is the social role expected from modern children as they leave the security of their home and family, and engage in wider social interactions, such as school.

The adolescence package is defined by a growth spurt and neuroendocrine changes leading to puberty and adolescent behaviors, as discussed earlier. It is assigned with determination of the age and length of fecundity. It entails plasticity in adapting to energy resources, other environmental cues, the social needs of adolescence and their maturation to determine fitness directly. Gluckman and Hanson proposed that human females evolved to enter puberty at a relatively young age and progressed to reproductive competence at 11–13 years of age (Gluckman and Hanson 2006). They argued that this would have matched the degree of psychosocial maturation necessary to function as an adult in Paleolithic hunter-gatherer societies. Juvenile preparation for adolescence would accordingly take place at ages 6–11.

Over the following agrarian period and modern civilization, they argue, biological puberty in females significantly precedes, rather than being matched to, the age of successful functioning as an adult. The disintegration of the adolescent package—the mismatch of biological and psychosocial maturation that has only appeared in the past 100 years—created fundamental pressures on contemporary adolescents and on how they live in society (Gluckman and Hanson 2006).

14

CONCLUDING REMARKS

This treatise has been an attempt to use life-history theory in understanding child growth in a broad evolutionary perspective. Life-history traits respond to environmental cues not to improve our health, but rather in order to enhance fecundity-survival schedules and behavioral strategies that yield the highest reproductive fitness in a given environment. It attempts to explain how humans evolved a gestation that was lengthy, but not long enough; large-bodied, yet immature, newborns; an unusually high rate of energy-costly postnatal brain growth; and an extended period of offspring dependency and slow growth in which the young are helpless. It also endeavors to explain a brief duration of breast-feeding, but intense maternal, familial, and quality care of the child, a matchless adolescence with concealed ovulation, delayed reproduction, and a menopause.

Like other organisms, humans evolved to withstand environmental hardships like energy crises by responding in ways that would maintain evolutionary fitness, even if submaximal. The means to do this is a series of predictive adaptive responses that utilize the sensitive times of transitions from one to its next life-history stage, each assigned with its own domain (Hochberg 2009) (Fig. 14.1). The transition from infancy to childhood is assigned with predictive adaptation to the environment of body size. Short-term adaptations to energy crises utilize a plasticity that postpones the timing of transition to reduce height and advances timing to enhance size. The transition from childhood to juvenility is part of a strategy in the transition from a period of total dependence on the family and tribe for provision and protection into self-supply; it is assigned with a predictive adaptive response of body composition and energy metabolism. The transition from juvenility to adolescence is assigned with longevity, when food supply is short, and with the age and length of fecundity

Evo-Devo of Child Growth: Treatise on Child Growth and Human Evolution, First Edition.
Ze'ev Hochberg.
© 2012 Wiley-Blackwell. Published 2012 by John Wiley & Sons, Inc.

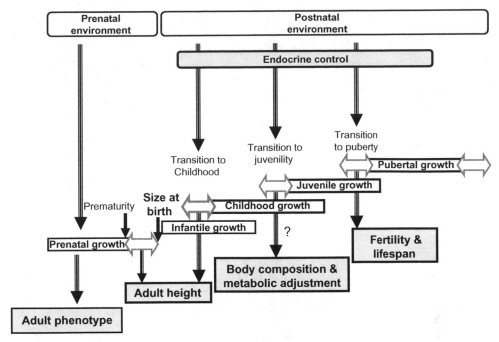

Fig. 14.1. Periods of adaptive plasticity in the transition between life-history stages (double arrows). Transition from infancy to childhood couples with a predictive adaptive response that determines adult height. Transition from childhood to juvenility bestows a predictive adaptive response that resolves adult body composition and metabolic consequences. The transition from juvenility to adolescence establishes the life span, age of fecundity, and reproduction. Adapted with permission from Hochberg, Feil et al. (2011).

and fertility. It entails plasticity in adapting to energy resources, other environmental cues, the social needs of adolescence, and their maturation to determine fitness directly.

These periods influence each other in an intricate web of connections that are related to evolutionary fitness and lifelong advantages. Human body size is an important paradigm for understanding the relationship between phenotypic adaptive plasticity and adaptive genetic changes. Whereas the gradual tendency to be bigger over the last 4 million years is genetic, the gradual secular trend over the last 150 years is not, as the short time interval does not provide the necessary timescale.

The implications of disintegration of this scheme in industrial societies of the 20th century and beyond are not much appreciated, but opt to result in conflicts that start to be seen. Among them is the obesity epidemic, the PCOS epidemic, and earlier pubertal development of girls. I have dwelt considerably on the physical, mental, and social interactions of each life-history stage. Their disintegration will also influence mental and social behaviors.

Human evolution has selected much of its preference by a special credence given to brain development. It is remarkable that the nervous system not only regulates most of the existing functions of the body, including perseverance, strength, and body size, but has indirectly influenced the progressive development of various bodily structures and of certain mental qualities through the exertion of choice by sexual selection, social behaviors, and culture. Powers of the mind, the influence of love and jealousy, and appreciation of the beautiful clearly depend on evolution of the brain. Even so, it is fascinating that humans, birds, fish, and butterflies have a similar perception of beauty, as their females' preference for good-looking males are comparable.

In the past decade we speak of translational research as a way of thinking about and conducting scientific research to make the results of basic research applicable to the population. In the field of medicine, for example, it is used to translate the findings in basic research more quickly and efficiently into medical practice and, thus, meaningful health outcomes, whether those are physical, mental, or social outcomes. Translational research has come to be seen as the key missing research component to improve medical care. Traditionally, translational research has followed a one-dimensional sequence of events: discovery of a new mechanism and a potential target in an experimental or basic research setting, development of a biomarker, validation of its utility in the clinical setting, and its eventual introduction into clinical practice. Translational evolution-based research in child health must be seen as a reiterative process that ranges from research in basic evolution, preclinical research, anthropology, and pediatric clinical research.

Hereditary, environmental, and stochastic factors determine child growth in his unique environment, but their relative contribution to the phenotypic outcome and the extent of stochastic epigenetic reprogramming that is required to alter human phenotypes is not known because few data are available (Hochberg, Feil et al. 2011). If the environment can influence growth and developmental trajectories during youth life-history stages and later life outcomes, how do epigenetic events influence the transition from one life-history stage to the next, growth, and puberty at the molecular level? Growth and puberty are regulated by insulin, growth hormone, the IGFs, and the sex hormones, to mention a few of the control hormones. These hormones drive the rate of growth and development, but it is unclear how the environment determines the timing of the different phases of developmental events and the quantity of growth.

Epigenetic mechanisms potentially play an important role in the "developmental origins of health and adult disease" phenomenon. Environmental influences during embryonic and early life development can permanently alter epigenetic gene regulation, which in turn can result in imprinting and reprogramming of the epigenome, and influence child growth, maturation, development, and body composition in later life-history stages. The mechanisms by which cues about nutrient availability in the uterus and postnatal environment are transmitted to the offspring and by which different stable phenotypes are induced are still unknown. The genetic control of the regulation of placental supply and fetal demand for maternal nutrients is not fully understood, and many of the detrimental events that occur in the fetus, infant, and later youth stages could be possibly due to epigenetic misprogramming. Epigenetic transgenerational traits are a new paradigm in child growth and development that has not been considered previously. In fact, research into epigenetic transgenerational disease is now one of the new topics that is undertaken in order to understand the etiology of disease states that have a familial inheritance, but do not follow normal genetic mechanisms, with child growth as its most striking trait.

Since no other animal has a similar youth life history to that of humans, an obvious question is whether the findings from any experimental animal can be extrapolated to humans. Many tissues can be sampled in humans by noninvasive and minimally invasive methods: red blood cells, T cells, sperm, placentae, umbilical tissue and blood, fetal cells and/or fetal DNA in the maternal circulation, foreskins, urine, cord blood, nails, and hair. Obviously, the lineage of the specimens is highly important when studying epigenetic mechanisms. For example, foreskin derives from endoderm, and cord blood from the mesoderm. Appropriate specimen collection includes the need for robust protocols for sample preparation, storage, and retrieval. In addition, accurate phenotyping of the donor is crucial in order to ensure the reliability of the data from any past, current, or future cohort studies. Finally, the available options for noninvasive sample collection overcome some, but not all, of the stringent ethical requirements for conducting experiments in humans.

Focusing specifically on the needs and opportunities in child health, we need better phenotypic assessments than those we use currently in order to define study populations, and in particular to distinguish intrauterine growth restriction (IUGR) infants from other SGA infants. The SGA infant embraces two different phenotypes: (1) a small infant who has been a small fetus throughout pregnancy with a normal fetal growth rate, and (2) an IUGR infant with reduced fetal growth rate. In order to distinguish between these two phenotypes, an accurate classification for identifying each phenotype is critical for the clinician in order to (a) reduce avoidable perinatal morbidity and mortality, and (b) more accurately assess the risk of developing disease in later life. Although birth weight and length are easily obtainable, they are inadequate indices to use to fully phenotype SGA and IUGR infants, even with additional information on ethnicity, sex, and parental size. Birth weight and length are sometimes crudely used as indicators of fetal growth and nutrition, but these are measures of attained size, rather than measures of fetal growth rate. Epigenetic biomarkers have the potential to greatly improve the phenotyping of these subsets of infants, and this is an example of the need for such a biomarker in child health.

Perhaps the most fundamental set of questions raised by the life-history approach to child growth concerns how the unique growth pattern of each child in his given genetic background and current environment best serves his reproductive fitness. Further research is required to understand the mechanisms underlying the energy allocation process toward growth during youth stages. Another direction will be to understand how these mechanisms interact with socioeconomic conditions in generating behaviors that affect life history and growth. The obesity epidemics that affect us in recent decades require special attention for evolution under unlimited resources.

Charles Darwin, in *The Descent of Man, and Selection in Relation to Sex*, Vol. 2, stated: "The great principle of evolution stands up clear and firm, when these groups or facts are considered in connection with others. The grounds upon which these conclusions rest will not be shaken; they have long been known." But, until recently, they told us nothing with respect to child growth and body size. Now, when viewed in light of our knowledge of life-history theory, life-history stages, and the transitions between them, their meanings take a new turn.

REFERENCES

AAP. 2000. "American Academy of Pediatrics. Committee on Nutrition. Hypoallergenic infant formulas." *Pediatrics* 106(2 Pt 1): 346–349.

Agostoni, C., and C. Braegger et al. 2009. "Breast-feeding: A commentary by the ESPGHAN Committee on Nutrition." *J Pediatr Gastroenterol Nutr* 49(1): 112–125.

Aimone, J.B., W. Deng et al. 2010. "Adult neurogenesis: Integrating theories and separating functions." *Trends Cogn Sci* 14(7): 325–337.

Al-Attas, O.S., N.M. Al-Daghri, M. Alokail et al. 2010. "Adiposity and insulin resistance correlate with telomere length in middle-aged Arabs: The influence of circulating adiponectin." *Eur J Endocrinol* 163(4): 601–607.

Allal, N., R. Sear et al. 2004. "An evolutionary model of stature, age at first birth and reproductive success in Gambian women." *Proc Biol Sci* 271(1538): 465–470.

Allen, H.L., K. Estrada, G. Lettre et al. 2010. "Hundreds of variants clustered in genomic loci and biological pathways affect human height." *Nature* 467(7317): 832–838.

Alper, J. 2000. "New insights into type 2 diabetes." *Science* 289: 37–39.

Altman, J., and G.D. Das. 1965. "Autoradiographic and histological evidence of postnatal hippocampal neurogenesis in rats." *J Comp Neurol* 124(3): 319–335.

Ambros, V., and H.R. Horvitz. 1984. "Heterochronic mutants of the nematode Caenorhabditis elegans." *Science* 226(4673): 409–416.

Anderson, J., D. Viskochil et al. 2002. "Gastrointestinal complications of Russell-Silver syndrome: A pilot study." *Am J Med Genet* 113(1): 15–19.

Androutsellis-Theotokis, A., R.R. Leker et al. 2006. "Notch signalling regulates stem cell numbers *in vitro* and *in vivo*." *Nature* 442(7104): 823–826.

Androutsellis-Theotokis, A., M.A. Rueger et al. 2008. "Signaling pathways controlling neural stem cells slow progressive brain disease." *Cold Spring Harb Symp Quant Biol* 73: 403–410.

Evo-Devo of Child Growth: Treatise on Child Growth and Human Evolution, First Edition.
Ze'ev Hochberg.
© 2012 Wiley-Blackwell. Published 2012 by John Wiley & Sons, Inc.

Androutsellis-Theotokis, A., M.A. Rueger et al. 2009. "Targeting neural precursors in the adult brain rescues injured dopamine neurons." *Proc Natl Acad Sci USA* 106(32): 13570–13575.

Androutsellis-Theotokis, A., M.A. Rueger, D.M. Park et al. 2010. "Angiogenic factors stimulate growth of adult neural stem cells." *PLoS One* 5(2): e9414.

Angrisano, T., F. Lembo, R. Pero et al. 2006. "TACC3 mediates the association of MBD2 with histone acetyltransferases and relieves transcriptional repression of methylated promoters." *Nucleic Acids Res* 34(1): 364–372.

Ankarberg Lindgren, C. 2005. "Testosterone and 17beta-oestradiol secretion in children and adolescents. Assay development, levels for comparison and clinical applications." PhD dissertation, Goteborg University.

Anway, M.D., A.S. Cupp et al. 2005. "Epigenetic transgenerational actions of endocrine disruptors and male fertility." *Science* 308(5727): 1466–1469.

Arcaleni, E. 2006. "Secular trend and regional differences in the stature of Italians, 1854–1980." *Econ Hum Biol* 4(1): 24–38.

Arlt, W., F. Callies et al. 2000. "DHEA replacement in women with adrenal insufficiency—pharmacokinetics, bioconversion and clinical effects on well-being, sexuality and cognition." *Endocr Res* 26(4): 505–511.

Arlt, W., J.W. Martens et al. 2002. "Molecular evolution of adrenarche: Structural and functional analysis of p450c17 from four primate species." *Endocrinology* 143(12): 4665–4672.

Arquitt, A.B., B.J. Stoecker et al. 1991. "Dehydroepiandrosterone sulfate, cholesterol, hemoglobin, and anthropometric measures related to growth in male adolescents." *J Am Diet Assoc* 91(5): 575–579.

Arvidsson, A., T. Collin et al. 2002. "Neuronal replacement from endogenous precursors in the adult brain after stroke." *Nat Med* 8(9): 963–970.

Auchus, R.J., and W.E. Rainey. 2004. "Adrenarche—physiology, biochemistry and human disease." *Clin Endocrinol (Oxf)* 60(3): 288–296.

Baird, J., D. Fisher, P. Lucas et al. 2005. "Being big or growing fast: Systematic review of size and growth in infancy and later obesity." *BMJ* 331(7522): 929.

Baquedano, M.S., E. Berensztein et al. 2005. "Expression of the IGF system in human adrenal tissues from early infancy to late puberty: Implications for the development of adrenarche." *Pediatr Res* 58(3): 451–458.

Barker, D.J. 1992a. "Fetal growth and adult disease." *Br J Obstet Gynaecol* 99(4): 275–276.

Barker, D.J. 1992b. "The fetal origins of adult hypertension." *J Hypertens Suppl* 10(7): S39–44.

Barker, D.J. 1992c. "The fetal origins of diseases of old age." *Eur J Clin Nutr* 46(Suppl 3): S3–9.

Barker, D.J. 1995. "Fetal origins of coronary heart disease." *BMJ* 311(6998): 171–174.

Barker, D.J. 2006. "Adult consequences of fetal growth restriction." *Clin Obstet Gynecol* 49(2): 270–283.

Barker, D.J., and C. Osmond. 1986. "Infant mortality, childhood nutrition, and ischaemic heart disease in England and Wales." *Lancet* 1(8489): 1077–1081.

Barker, D.J., C. Osmond et al. 1989. "The intrauterine and early postnatal origins of cardiovascular disease and chronic bronchitis." *J Epidemiol Community Health* 43(3): 237–240.

Barreto, G., A. Schafer, J. Marhold et al. 2007. "Gadd45a promotes epigenetic gene activation by repair-mediated DNA demethylation." *Nature* 445(7128): 671–675.

Bartlett, P.F., H.H. Reid et al. 1988. "Immortalization of mouse neural precursor cells by the c-myc oncogene." *Proc Natl Acad Sci USA* 85(9): 3255–3259.

Bateson, P. 2007. "Developmental plasticity and evolutionary biology." *J Nutr* 137(4): 1060–1062.

Bateson, P., D. Barker et al. 2004. "Developmental plasticity and human health." *Nature* 430(6998): 419–421.

Beardsall, K., K.K. Ong, N. Murphy et al. 2009. "Heritability of childhood weight gain from birth and risk markers for adult metabolic disease in prepubertal twins." *J Clin Endocrinol Metab* 94(10): 3708–3713.

Bell, J.T., N.J. Timpson, N.W. Rayner et al. 2011. "Genome-wide association scan allowing for epistasis in type 2 diabetes." *Annals of Human Genetics* 75: 10–19.

Bellis, M.A., J. Downing et al. 2006. "Adults at 12? Trends in puberty and their public health consequences." *J Epidemiol Community Health* 60(11): 910–911.

Belsky, J., and R.M. Fearon. 2002. "Early attachment security, subsequent maternal sensitivity, and later child development: Does continuity in development depend upon continuity of caregiving?" *Attach Hum Dev* 4(3): 361–387.

Belsky, J., L. Steinberg, and P. Draper. 1991. "Childhood experience, interpersonal development, and reproductive strategy: An evolutionary theory of socialization." *Child Dev* 62(4): 647–670.

Ben-Nathan, D., S. Lustig et al. 1992. "Dehydroepiandrosterone protects mice inoculated with West Nile virus and exposed to cold stress." *J Med Virol* 38(3): 159–166.

Berkey, C.S., J.D. Gardner et al. 2000. "Relation of childhood diet and body size to menarche and adolescent growth in girls." *Am J Epidemiol* 152(5): 446–452.

Bird, A. 2002. "DNA methylation patterns and epigenetic memory." *Genes Dev* 16(1): 6–21.

Bjorkqvist, M., C. Dornonville de la Cour et al. 2002. "Role of gastrin in the development of gastric mucosa, ECL cells and A-like cells in newborn and young rats." *Regul Pept* 108(2–3): 73–82.

Black, R.E., S.S. Morris et al. 2003. "Where and why are 10 million children dying every year?" *Lancet* 361(9376): 2226–2234.

Blogowska, A., I. Rzepka-Gorska et al. 2005. "Body composition, dehydroepiandrosterone sulfate and leptin concentrations in girls approaching menarche." *J Pediatr Endocrinol Metab* 18(10): 975–983.

Blurton Jones, N. 1987. "Bushman birth spacing: Direct tests of some simple predictions." *Ethol Sociobiol* 8: 183–203.

Blurton Jones, N.G., K. Hawkes et al. 2002. "Antiquity of postreproductive life: Are there modern impacts on hunter-gatherer postreproductive life spans?" *Am J Hum Biol* 14(2): 184–205.

Blurton Jones, N.G., L.C. Smith et al. 1992. "Demography of the Hadza, an increasing and high density population of Savanna foragers." *Am J Phys Anthropol* 89(2): 159–181.

Bogin, B. 1999a. "Evolutionary perspective on human growth." *Annu Rev Anthropol* 28: 109–153.

Bogin, B. 1999b. *Patterns of Human Growth*. Cambridge: Cambridge University Press.

Bogin, B. 2002. "The evolution of human growth." *Human Growth and Development*, edited by N. Cameron, 295–320. Amsterdam: Academic Press.

Bogin, B., M.I. Silva, and L. Rios. 2007. "Life history trade-offs in human growth: Adaptation or pathology?" *Am J Hum Biol* 19: 631–642.

Bornstein, S. 2006. *Evolution, Stress and Modern Medicine*. Dresden: Progressmedia Verlag und Werbeagentur.

Bornstein, S.R., A. Schuppenies et al. 2006. "Approaching the shared biology of obesity and depression: The stress axis as the locus of gene-environment interactions." *Mol Psychiatry* 11(10): 892–902.

Bostick, M., J.K. Kim, P.O. Esteve et al. 2007. "UHRF1 plays a role in maintaining DNA methylation in mammalian cells." *Science* 317(5845): 1760–1764.

Bourc'his, D., and T.H. Bestor. 2004. "Meiotic catastrophe and retrotransposon reactivation in male germ cells lacking Dnmt3L." *Nature* 431(7004): 96–99.

Bourc'his, D., G.L. Xu, C.S. Lin et al. 2001. "Dnmt3L and the establishment of maternal genomic imprints." *Science* 294(5551): 2536–2539.

Bouret, S.G., S.J. Draper et al. 2004. "Trophic action of leptin on hypothalamic neurons that regulate feeding." *Science* 304(5667): 108–110.

Bourguignon, J.P. 1988. "Linear growth as a function of age at onset of puberty and sex steroid dosage: Therapeutic implications." *Endocr Rev* 9(4): 467–488.

Bowman, J.E. and P.C. Lee. 1995. "Growth and threshold weaning weights among captive rhesus macaques." *Am J Phys Anthropol* 96(2): 159–175.

Brawley, L., C. Torrens et al. 2004. "Glycine rectifies vascular dysfunction induced by dietary protein imbalance during pregnancy." *J Physiol* 554(Pt 2): 497–504.

Brown, J.H., and J.F. Gillooly. 2003. "Ecological food webs: High-quality data facilitate theoretical unification." *Proc Natl Acad Sci USA* 100(4): 1467–1468.

Brown, P., T. Sutikna et al. 2004. "A new small-bodied hominin from the Late Pleistocene of Flores, Indonesia." *Nature* 431(7012): 1055–1061.

Brown, S.E., and M. Szyf. 2007. "Epigenetic programming of the rRNA promoter by MBD3." *Mol Cell Biol* 27(13): 4938–4952.

Bruel-Jungerman, E., S. Laroche et al. 2005. "New neurons in the dentate gyrus are involved in the expression of enhanced long-term memory following environmental enrichment." *Eur J Neurosci* 21(2): 513–521.

Bruniquel, D., and R.H. Schwartz. 2003. "Selective, stable demethylation of the interleukin-2 gene enhances transcription by an active process." *Nat Immunol* 4(3): 235–240.

Bull, N.D., and P.F. Bartlett. 2005. "The adult mouse hippocampal progenitor is neurogenic but not a stem cell." *J Neurosci* 25(47): 10815–10821.

Burdge, G.C., K.A. Lillycrop et al. 2008. "The nature of the growth pattern and of the metabolic response to fasting in the rat are dependent upon the dietary protein and folic acid intakes of their pregnant dams and post-weaning fat consumption." *Br J Nutr* 99(3): 540–549.

Burdge, G.C., K.A. Lillycrop et al. 2009. "Folic acid supplementation during the juvenile-pubertal period in rats modifies the phenotype and epigenotype induced by prenatal nutrition." *J Nutr* 139(6): 1054–1060.

Burdge, G.C., J. Slater-Jefferies et al. 2007. "Dietary protein restriction of pregnant rats in the F0 generation induces altered methylation of hepatic gene promoters in the adult male offspring in the F1 and F2 generations." *Br J Nutr* 97(3): 435–439.

Burger, H.G., Y. Yamada et al. 1991. "Serum gonadotropin, sex steroid, and immunoreactive inhibin levels in the first two years of life." *J Clin Endocrinol Metab* 72(3): 682–686.

Cameron, H.A., C.S. Woolley et al. 1993. "Differentiation of newly born neurons and glia in the dentate gyrus of the adult rat." *Neuroscience* 56(2): 337–344.

Campbell, B. 2006. "Adrenarche and the evolution of human life history." *Am J Hum Biol* 18(5): 569–589.

Campbell, J.H., and P. Perkins. 1988. "Transgenerational effects of drug and hormonal treatments in mammals: A review of observations and ideas." *Prog Brain Res* 73: 535–553.

Cano, B., M. de Oya, M. Benavente et al. 2006. "Dehydroepiandrosterone sulfate (DHEA-S) distribution in Spanish prepuberal children: Relationship with fasting plasma insulin concentrations and insulin resistance." *Clin Chim Acta* 366(1–2): 163–167.

Caspi, A., K. Sugden, T.E. Moffitt et al. 2003. "Influence of life stress on depression: Moderation by a polymorphism in the 5-HTT gene." *Science* 301(5631): 386–389.

Cattaneo, E., and R. McKay. 1990. "Proliferation and differentiation of neuronal stem cells regulated by nerve growth factor." *Nature* 347(6295): 762–765.

Cauffman, E., and L. Steinberg. 2000. "(Im)maturity of judgment in adolescence: Why adolescents may be less culpable than adults." *Behav Sci Law* 18(6): 741–760.

CDC. 2000. CDC Growth Charts: United States, http://www.cdc.gov/growthcharts/.

Chang, C.L., J.J. Cai, C. Lo et al. 2011. "Adaptive selection of an incretin gene in Eurasian populations." *Genome Research* 21: 21–32.

Charnov, E.L. 1991. "Evolution of life history variation among female mammals." *Proc Natl Acad Sci USA* 88(4): 1134–1137.

Chellakooty, M., I.M. Schmidt et al. 2003. "Inhibin A, inhibin B, follicle-stimulating hormone, luteinizing hormone, estradiol, and sex hormone-binding globulin levels in 473 healthy infant girls." *J Clin Endocrinol Metab* 88(8): 3515–3520.

Chen, C.C., and C.R. Parker, Jr. 2004. "Adrenal androgens and the immune system." *Semin Reprod Med* 22(4): 369–377.

Chen, Y.F., C.Y. Wu, C.H. Kao, and T.F. Tsai. 2010. "Longevity and lifespan control in mammals: Lessons from the mouse." *Ageing Res Rev* 9(Suppl 1): S28–S35.

Chen-Pan, C., I.J. Pan et al. 2002. "Recovery of injured adrenal medulla by differentiation of pre-existing undifferentiated chromaffin cells." *Toxicol Pathol* 30(2): 165–172.

Christakis, N.A., and J.H. Fowler. 2007. "The spread of obesity in a large social network over 32 years." *N Engl J Med* 357(4): 370–379.

Chrousos, G.P. 1995. "The hypothalamic–pituitary–adrenal axis and immune-mediated inflammation." *N Engl J Med* 332: 1351–1362.

Chrousos, G.P. 1998. " Stressors, stress and neuroendocrine integration of the adaptive response: 1997 Hans Selye Memorial Lecture." *Ann NY Acad Sci* 851: 311–335.

Chrousos G.P. 2000. "The stress response and immune function: Clinical implications: The 1999 Novera H. Spector Lecture." *Ann NY Acad Sci* 917: 38–67.

Chrousos, G.P. 2004. "The glucocorticoid receptor gene, longevity, and the highly prevalent complex disorders of western societies." *Am J Medicine* 117: 204–207.

Chrousos, G.P. 2007. "Organization and integration of the endocrine system: The sleep and wakefulness perspective." *Sleep Medicine Clinics* 2: 125–145.

Chrousos, G.P. 2009. "Stress and disorders of the stress system." *Nature Endocrinology Reviews* 5(7): 374–381.

Chrousos, G.P., and P.W. Gold. 1992. "The concepts of stress and stress system disorders: Overview of physical and behavioral homeostasis." *JAMA* 267: 1244–1252.

Chrousos, G.P., and T. Kino. 2009. "Glucocorticoid signaling in the cell." *Ann NY Acad Sci* 1179: 153–166.

Chrousos, G.P., D.L. Loriaux, P.W. Gold, eds. 1988. "Mechanisms of physical and emotional stress." In: *Advances in Experimental Medicine and Biology*, Vol. 245. New York: Plenum Press.

Chung, K.F., F. Sicard et al. 2009. "Isolation of neural crest derived chromaffin progenitors from adult adrenal medulla." *Stem Cells* 27(10): 2602–2613.

Cohen, J.E., T. Jonsson et al. 2003. "Ecological community description using the food web, species abundance, and body size." *Proc Natl Acad Sci USA* 100(4): 1781–1786.

Cole, T.J., J.V. Freeman et al. 1995. "Body mass index reference curves for the UK, 1990." *Arch Dis Child* 73(1): 25–29.

Comb, M., and H.M. Goodman. 1990. "CpG methylation inhibits proenkephalin gene expression and binding of the transcription factor AP-2." *Nucleic Acids Res* 18(13): 3975–3982.

Coppa, A., F. Manni, C. Stringer et al. 2007. "Evidence for new Neanderthal teeth in Tabun Cave (Israel) by the application of self-organizing maps (SOMs)." *J Hum Evol* 52(6): 601–613.

Corbo, R.M., and R. Scacchi. 1999. "Apolipoprotein E (APOE) allele distribution in the world. Is *APOE*4* a 'thrifty' allele?" *Ann Hum Genet* 63: 301–310.

Craig, C.G., V. Tropepe et al. 1996. "*In vivo* growth factor expansion of endogenous subependymal neural precursor cell populations in the adult mouse brain." *J Neurosci* 16(8): 2649–2658.

Crespi, E.J., and R.J. Denver. 2005. "Ancient origins of human developmental plasticity." *Am J Hum Biol* 17(1): 44–54.

Crofton, P.M., A.E. Evans et al. 2002. "Inhibin B in boys from birth to adulthood: Relationship with age, pubertal stage, FSH and testosterone." *Clin Endocrinol (Oxf)* 56(2): 215–221.

Cronk, C., A.C. Crocker et al. 1988. "Growth charts for children with Down syndrome: 1 month to 18 years of age." *Pediatrics* 81(1): 102–110.

Crow, J.F. 1961. "Population genetics." *Am J Hum Genet* 13: 137–150.

Cumming, D.C., R.W. Rebar et al. 1982. "Evidence for an influence of the ovary on circulating dehydroepiandrosterone sulfate levels." *J Clin Endocrinol Metab* 54(5): 1069–1071.

Cummins, A.G., and F.M. Thompson. 2002. "Effect of breast milk and weaning on epithelial growth of the small intestine in humans." *Gut* 51(5): 748–754.

Cunnane, S.C. 2005. *Survival of the Fattest: The Key to Human Brain Evolution*. Singapore: World Scientific Publishing Co.

Curtis, M.A., E.B. Penney et al. 2003. "Increased cell proliferation and neurogenesis in the adult human Huntington's disease brain." *Proc Natl Acad Sci USA* 100(15): 9023–9027.

Cutler, G.B. Jr., M. Glenn et al. 1978. "Adrenarche: A survey of rodents, domestic animals, and primates." *Endocrinology* 103(6): 2112–2118.

D'Alessio, A.C., and M. Szyf. 2006. "Epigenetic tete-a-tete: The bilateral relationship between chromatin modifications and DNA methylation." *Biochem Cell Biol* 84(4): 463–476.

Dabiri, J.O., S.P. Colin et al. 2007. "Morphological diversity of medusan lineages constrained by animal-fluid interactions." *J Exp Biol* 210(Pt 11): 1868–1873.

Dallman, M.F., N. Pecoraro et al. 2003. "Chronic stress and obesity: A new view of 'comfort food.' " *Proc Natl Acad Sci USA* 100(20): 11696–11701.

Dang, S., H. Yan et al. 2004. "Poor nutritional status of younger Tibetan children living at high altitudes." *Eur J Clin Nutr* 58(6): 938–946.

Dasenbrock, C., T. Tillmann et al. 2005. "Maternal effects and cancer risk in the progeny of mice exposed to X-rays before conception." *Exp Toxicol Pathol* 56(6): 351–360.

De Zegher, F., M. Bettendorf et al. 1988. "Hormone ontogeny in the ovine fetus: XXI. The effect of insulin-like growth factor-I on plasma fetal growth hormone, insulin and glucose concentrations." *Endocrinology* 123(1): 658–660.

De Zegher, F., J. Kimpen et al. 1990. "Hypersomatotropism in the dysmature infant at term and preterm birth." *Biol Neonate* 58(4): 188–191.

Dean, C. 2007. "Growing up slowly 160,000 years ago." *Proc Natl Acad Sci USA* 104(15): 6093–6094.

Dechert, U., A.M. Duncan et al. 1995. "Protein-tyrosine phosphatase SH-PTP2 (PTPN11) is localized to 12q24.1-24.3." *Hum Genet* 96(5): 609–615.

Del Giudice, M. 2009. "Sex, attachment, and the development of reproductive strategies." *Behav Brain Sci* 32(1): 1–21; discussion 21–67.

Deloulme, J.C., C. Gensburger et al. 1991. "Effects of basic fibroblast growth factor on the development of GABAergic neurons in culture." *Neuroscience* 42(2): 561–568.

Demerath, E.W., A.C. Choh, S.A. Czerwinski et al. 2007. "Genetic and environmental influences on infant weight and weight change: The Fels Longitudinal Study." *Am J Hum Biol* 19(5): 692–702.

Detich, N., J. Theberge, and M. Szyf. 2002. "Promoter-specific activation and demethylation by MBD2/demethylase." *J Biol Chem* 277(39): 35791–35794.

Diamond, J. 2003. "The double puzzle of diabetes." *Nature* 423: 599–602.

Doetsch, F., I. Caille et al. 1999. "Subventricular zone astrocytes are neural stem cells in the adult mammalian brain." *Cell* 97(6): 703–716.

Dolinoy, D.C., R. Das, J.R. Weidman et al. 2007. "Metastable epialleles, imprinting, and the fetal origins of adult diseases." *Pediatr Res* 61(5 Pt 2): 30R–37R.

Dolinoy, D.C., D. Huang, R.L. Jirtle et al. 2007. "Maternal nutrient supplementation counteracts bisphenol A-induced DNA hypomethylation in early development." *Proc Natl Acad Sci USA* 104(32): 13056–13061.

Dolinoy, D.C., and R.L. Jirtle. 2008. "Environmental epigenomics in human health and disease." *Environ Mol Mutagen* 49(1): 4–8.

Dolinoy, D.C., J.R. Weidman, R.A. Waterland et al. 2006. "Maternal genistein alters coat color and protects Avy mouse offspring from obesity by modifying the fetal epigenome." *Environ Health Perspect* 114(4): 567–572.

Dong, E., M. Nelson, D.R. Grayson et al. 2008. "Clozapine and sulpiride but not haloperidol or olanzapine activate brain DNA demethylation." *Proc Natl Acad Sci USA* 105(36): 13614–13619.

Dorn, L.D., S.F. Hitt et al. 1999. "Biopsychological and cognitive differences in children with premature vs. on-time adrenarche." *Arch Pediatr Adolesc Med* 153(2): 137–146.

Drewett, R.F., S.S. Corbett et al. 1999. "Cognitive and educational attainments at school age of children who failed to thrive in infancy: A population-based study." *J Child Psychol Psychiatry* 40(4): 551–561.

Drooger, J.C., J.W. Troe et al. 2005. "Ethnic differences in prenatal growth and the association with maternal and fetal characteristics." *Ultrasound Obstet Gynecol* 26(2): 115–122.

Dunger, D.B., M.L. Ahmed et al. 2006. "Early and late weight gain and the timing of puberty." *Mol Cell Endocrinol* 254–255: 140–145.

Egger, G., G. Liang et al. 2004. "Epigenetics in human disease and prospects for epigenetic therapy." *Nature* 429(6990): 457–463.

Ehrhart-Bornstein, M., and S.R. Bornstein. 2008. "Cross-talk between adrenal medulla and adrenal cortex in stress." *Ann NY Acad Sci* 1148: 112–117.

Elks, C.E., R.J. Loos, S.J. Sharp et al. 2010. "Genetic markers of adult obesity risk are associated with greater early infancy weight gain and growth." *PLoS Medicine* 7(5): e1000284.

Elks, C.E., J.R. Perry, P. Sulem et al. 2010. "Thirty new loci for age at menarche identified by a meta-analysis of genome-wide association studies." *Nat Genet* 42: 1077–1085.

Ellegren, H. 2000. "Microsatellite mutations in the germline: Implications for evolutionary inference." *Trends Genet* 16(12): 551–558.

Ellis, K.J. 1997. "Body composition of a young, multiethnic, male population." *Am J Clin Nutr* 66(6): 1323–1331.

Ellis, K.J., S.A. Abrams et al. 1997. "Body composition of a young, multiethnic female population." *Am J Clin Nutr* 65(3): 724–731.

Ellison, P.T. 1981. "Morbidity, morality, and menarche." *Hum Biol* 53(4): 635–643.

Ernst, C., V. Deleva, X. Deng et al. 2009. "Alternative splicing, methylation state, and expression profile of tropomyosin-related kinase B in the frontal cortex of suicide completers." *Arch Gen Psychiatry* 66(1): 22–32.

Even, L., A. Cohen et al. 2000. "Longitudinal analysis of growth over the first 3 years of life in Turner's syndrome." *J Pediatr* 137(4): 460–464.

Fak, F., L. Friis-Hansen et al. 2007. "Gastric ghrelin cell development is hampered and plasma ghrelin is reduced by delayed weaning in rats." *J Endocrinol* 192(2): 345–352.

Fall, C.H., H.S. Sachdev, C. Osmond et al. 2008. "Adult metabolic syndrome and impaired glucose tolerance are associated with different patterns of BMI gain during infancy: Data from the New Delhi Birth Cohort." *Diabetes Care* 31(12): 2349–2356.

Faruque, A.S., A.M. Ahmed et al. 2008. "Nutrition: Basis for healthy children and mothers in Bangladesh." *J Health Popul Nutr* 26(3): 325–339.

Feng, J., and G. Fan. 2009. "The role of DNA methylation in the central nervous system and neuropsychiatric disorders." *Int Rev Neurobiol* 89: 67–84.

Feng, J., Y. Zhou, S.L. Campbell et al. 2010. "Dnmt1 and Dnmt3a maintain DNA methylation and regulate synaptic function in adult forebrain neurons." *Nat Neurosci* 13(4): 423–430.

Flanagan, J.M., V. Popendikyte, N. Pozdniakovaite et al. 2006. "Intra- and interindividual epigenetic variation in human germ cells." *Am J Hum Genet* 79(1): 67–84.

Fleenor, D., J. Oden et al. 2005. "Roles of the lactogens and somatogens in perinatal and postnatal metabolism and growth: Studies of a novel mouse model combining lactogen resistance and growth hormone deficiency." *Endocrinology* 146(1): 103–112.

Forsdahl, A. 1977. "Are poor living conditions in childhood and adolescence an important risk factor for arteriosclerotic heart disease?" *Br J Prev Soc Med* 31(2): 91–95.

Francis, D., J. Diorio, and M.J. Meaney. 1999. "Nongenomic transmission across generations of maternal behavior and stress responses in the rat." *Science* 286(5442): 1155–1158.

Franks, S., M.I. McCarthy et al. 2006. "Development of polycystic ovary syndrome: Involvement of genetic and environmental factors." *Int J Androl* 29(1): 278–285; discussion 286–290.

Frederiksen, K., and R.D. McKay. 1988. "Proliferation and differentiation of rat neuroepithelial precursor cells *in vivo*." *J Neurosci* 8(4): 1144–1151.

Frisch, R.E., and R. Revelle. 1970. "Height and weight at menarche and a hypothesis of critical body weights and adolescent events." *Science* 169(943): 397–399.

Frisch, R.E., R. Revelle et al. 1973. "Components of weight at menarche and the initiation of the adolescent growth spurt in girls: Estimated total water, lean body weight and fat." *Hum Biol* 45(3): 469–483.

Fujita, H., R. Fujii, S. Aratani et al. 2003. "Antithetic effects of MBD2a on gene regulation." *Mol Cell Biol* 23(8): 2645–2657.

Fuks, F., W.A. Burgers, A. Brehm et al. 2000. "DNA methyltransferase Dnmt1 associates with histone deacetylase activity." *Nat Genet* 24(1): 88–91.

Fuks, F., W.A. Burgers, N. Godin et al. 2001. "Dnmt3a binds deacetylases and is recruited by a sequence-specific repressor to silence transcription." *Embo J* 20(10): 2536–2544.

Fullerton, S.M., A. Bartoszewicz, G. Ybazeta et al. 2002. "Geographic and haplotype structure of candidate type 2 diabetes-susceptibility variants at the calpain-10 locus." *American Journal of Human Genetics* 70: 1096–1106.

Fuso, A., G. Ferraguti et al. 2010. "Early demethylation of non-CpG, CpC-rich, elements in the myogenin 5′-flanking region: A priming effect on the spreading of active demethylation." *Cell Cycle* 9(19): 3965–3976.

Fuso, A., V. Nicolia, R.A. Cavallaro et al. 2011. "DNA methylase and demethylase activities are modulated by one-carbon metabolism in Alzheimer's disease models." *J Nutr Biochem* 22(3): 242–251.

Gajdos, Z.K., J.L. Butler, K.D. Henderson et al. 2008. "Association studies of common variants in ten hypogonadotropic hypogonadism genes with age at menarche." *J Clin Endocrinol Metab* 93(11): 4290–4298.

Gale, C.R., M.K. Javaid et al. 2007. "Maternal size in pregnancy and body composition in children." *J Clin Endocrinol Metab* 92(10): 3904–3911.

Gallou-Kabani, C., A. Vige et al. 2007. "Lifelong circadian and epigenetic drifts in metabolic syndrome." *Epigenetics* 2(3): 137–146.

Garnier, D., K.B. Simondon et al. 2005. "Longitudinal estimates of puberty timing in Senegalese adolescent girls." *Am J Hum Biol* 17(6): 718–730.

Gartner, L.M., J. Morton et al. 2005. "Breastfeeding and the use of human milk." *Pediatrics* 115(2): 496–506.

Gaston, V., Y. Le Bouc et al. 2001. "Analysis of the methylation status of the KCNQ1OT and H19 genes in leukocyte DNA for the diagnosis and prognosis of Beckwith-Wiedemann syndrome." *Eur J Hum Genet* 9(6): 409–418.

Gat-Yablonski, G., T. Ben-Ari et al. 2004. "Leptin reverses the inhibitory effect of caloric restriction on longitudinal growth." *Endocrinology* 145(1): 343–350.

Gavan, J.A. 1953. "Growth and development of the chimpanzee; a longitudinal and comparative study." *Hum Biol* 25(2): 93–143.

Gawlik, A., R.S. Walker, and Z. Hochberg. 2011. "Impact of infancy duration on adult size in 22 subsistence-based societies." *Acta Paediatr*, doi: 10.1111/j.1651-2227.2011.02395.x [Epub ahead of print].

German, A., I. Peter, G. Livshits, and Z. Hochberg. 2009. "Genetic and environmental influences on the aget of the infancy-childhood transition (ICT)." *Hormone Research* 71: Abs 767.

Gerver, W., and R. de Bruin. 1996. *Pediatric Morphometrics: A Reference Manual*. Maastricht: Universitaire Pers Maastricht.

Gicquel, C., S. Rossignol et al. 2005. "Epimutation of the telomeric imprinting center region on chromosome 11p15 in Silver-Russell syndrome." *Nat Genet* 37(9): 1003–1007.

Gimble, J.M., S. Zvonic et al. 2006. "Playing with bone and fat." *J Cell Biochem* 98(2): 251–266.

Girard, J., T. Issad et al. 1993. "Influence of the weaning diet on the changes of glucose metabolism and of insulin sensitivity." *Proc Nutr Soc* 52(2): 325–333.

Gluckman, P.H., M. Hanson. 2005. *The Fetal Matrix: Evolution, Development, and Disease*. New York: Cambridge University Press.

Gluckman, P.D., and M.A. Hanson. 2006. "Evolution, development and timing of puberty." *Trends Endocrinol Metab* 17(1): 7–12.

Gluckman, P.D., and M.A. Hanson 2007a. "Developmental plasticity and human disease: Research directions." *J Intern Med* 261(5): 461–471.

Gluckman, P.D., and M.A. Hanson. 2007b. *Mismatch*. Oxford: Oxford University Press.

Gluckman, P.D., M.A. Hanson et al. 2008. "Effect of *in utero* and early-life conditions on adult health and disease." *N Engl J Med* 359(1): 61–73.

Gluckman, P.D., M.A. Hanson et al. 2009. "Towards a new developmental synthesis: Adaptive developmental plasticity and human disease." *Lancet* 373(9675): 1654–1657.

Godfrey, K.M., and D.J. Barker. 2001. "Fetal programming and adult health." *Public Health Nutr* 4(2B): 611–624.

Gogtay, N., J.N. Giedd et al. 2004. "Dynamic mapping of human cortical development during childhood through early adulthood." *Proc Natl Acad Sci USA* 101(21): 8174–8179.

Gold, P.W., K.E. Gabry, M.R. Yasuda, and G.P. Chrousos. 2002. "Divergent endocrine abnormalities in melancholic and atypical depression: Clinical and pathophysiologic implications." *Endocrinol Metab Clin North America* 31: 37–62.

Goll, M.G., F. Kirpekar, K.A. Maggert et al. 2006. "Methylation of tRNAAsp by the DNA methyltransferase homolog Dnmt2." *Science* 311(5759): 395–398.

Goodman, C.S., and B.C. Coughlin. 2000. "Introduction. The evolution of evo-devo biology." *Proc Natl Acad Sci USA* 97(9): 4424–4425.

Gorai, I., K. Tanaka et al. 2003. "Estrogen-metabolizing gene polymorphisms, but not estrogen receptor-alpha gene polymorphisms, are associated with the onset of menarche in healthy postmenopausal Japanese women." *J Clin Endocrinol Metab* 88(2): 799–803.

Gould, S. 1977. *Ontegeny and Phylogeny*. Cambridge: Belknap Press.

Gron, A.M. 1962. "Prediction of tooth emergence." *J Dent Res* 41: 573–585.

Gruenbaum, Y., H. Cedar, and A. Razin. 1982. "Substrate and sequence specificity of a eukaryotic DNA methylase." *Nature* 295(5850): 620–622.

Gruenbaum, Y., R. Stein, H. Cedar, and A. Razin. 1981. "Methylation of CpG sequences in eukaryotic DNA." *FEBS Lett* 124(1): 67–71.

Grumbach, M.M. 2005. "A window of opportunity: The diagnosis of gonadotropin deficiency in the male infant." *J Clin Endocrinol Metab* 90(5): 3122–3127.

Grumbach, M.M., and D.M. Styne. 2003. "Puberty: Ontogeny, neuroendocrinology, physiology, disorders." In: *Williams Textbook of Endocrinology*, edited by P.R. Larsen, H.M. Kronenberg, S. Melmed, and K.S. Polonsky, 1115–1301. Philadelphia, PA: Saunders.

Guercio, G., M.A. Rivarola et al. 2002. "Relationship between the GH/IGF-I axis, insulin sensitivity, and adrenal androgens in normal prepubertal and pubertal boys." *J Clin Endocrinol Metab* 87(3): 1162–1169.

Guercio, G., M.A. Rivarola et al. 2003. "Relationship between the growth hormone/insulin-like growth factor-I axis, insulin sensitivity, and adrenal androgens in normal prepubertal and pubertal girls." *J Clin Endocrinol Metab* 88(3): 1389–1393.

Guo, Y., Y. Chen, H. Ito et al. 2006. "Identification and characterization of lin-28 homolog B (LIN28B) in human hepatocellular carcinoma." *Gene* 384: 51–61.

Guven, A., P. Cinaz et al. 2005. "Are growth factors and leptin involved in the pathogenesis of premature adrenarche in girls?" *J Pediatr Endocrinol Metab* 18(8): 785–791.

Haig, D. 2010. "Transfers and transitions: Parent-offspring conflict, genomic imprinting, and the evolution of human life history." *Proc Natl Acad Sci USA* 107(Suppl 1): 1731–1735.

Halpern, C.T., J.R. Udry et al. 1998. "Monthly measures of salivary testosterone predict sexual activity in adolescent males." *Arch Sex Behav* 27(5): 445–465.

Hamada, Y., T. Udono, M. Teramoto, and T. Sugawara. 1996. "The growth pattern of chimpanzees: Somatic growth and reproductive maturation in Pan troglodytes." *Primates* 37: 279–295.

Hamill, P.V., T.A. Drizd et al. 1979. "Physical growth: National Center for Health Statistics percentiles." *Am J Clin Nutr* 32(3): 607–629.

Hamilton, W. 1963. "The evolution of altruistic behavior." *American Naturalist* 97: 354–356.

Hamm, S., G. Just, N. Lacoste et al. 2008. "On the mechanism of demethylation of 5-methylcytosine in DNA." *Bioorg Med Chem Lett* 18(3): 1046–1049.

Hardy, R., A.K. Wills, A. Wong et al. 2010. "Life course variations in the associations between FTO and MC4R gene variants and body size." *Hum Mol Genet* 19(3): 545–552.

Harris, E.E., and D. Meyer. 2006. "The molecular signature of selection underlying human adaptations." *Am J Phys Anthropol* 131: 89–130.

Harvey, P., and T. Clutton-Brock. 1985. "Life history variation in primates." *Evolution* 39: 559–581.

Hau, M. 2007. "Regulation of male traits by testosterone: Implications for the evolution of vertebrate life histories." *Bioessays* 29(2): 133–144.

Haumaitre, C., O. Lenoir et al. 2008. "Histone deacetylase inhibitors modify pancreatic cell fate determination and amplify endocrine progenitors." *Mol Cell Biol* 28(20): 6373–6383.

Hauner, H., and Z. Hochberg. 2002. "Endocrinology of adipose tissue." *Horm Metab Res* 34(11–12): 605–606.

Havelock, J.C., R.J. Auchus et al. 2004. "The rise in adrenal androgen biosynthesis: Adrenarche." *Semin Reprod Med* 22(4): 337–347.

Haviland, W. 1967. "Stature at Tikal, Guatemala: Implications for ancient Maya demography and social organization." *American Antiquity* 32: 316–331.

Hawkes, K. 2003. "Grandmothers and the evolution of human longevity." *Am J Hum Biol* 15(3): 380–400.

Haworth, C.M., S. Carnell, E.L. Meaburn et al. 2008. "Increasing heritability of BMI and stronger associations with the FTO gene over childhood." *Obesity (Silver Spring)* 16(12): 2663–2668.

Helgason, A., S. Palsson, G. Thorleifsson et al. 2007. "Refining the impact of TCF7L2 gene variants on type 2 diabetes and adaptive evolution." *Nature Genetics* 39: 218–225.

Hellman, A., and A. Chess. 2007. "Gene body-specific methylation on the active X chromosome." *Science* 315(5815): 1141–1143.

Helm, S. 1969. "Secular trend in tooth eruption: A comparative study of Danish school children of 1913 and 1965." *Arch Oral Biol* 14(10): 1177–1191.

Hermanussen, M., A.P. Garcia et al. 2006. "Obesity, voracity, and short stature: The impact of glutamate on the regulation of appetite." *Eur J Clin Nutr* 60(1): 25–31.

Hertzman, C., C. Power, S. Matthews, and O. Manor. 2001. "Using an interactive framework of society and lifecourse to explain self-rated health in early adulthood." *Soc Sci Med* 53(12): 1575–1585.

Hickey, T.E., R.S. Legro et al. 2006. "Epigenetic modification of the X chromosome influences susceptibility to polycystic ovary syndrome." *J Clin Endocrinol Metab* 91(7): 2789–2791.

Hileman, S.M., D.D. Pierroz et al. 2000. "Leptin, nutrition, and reproduction: Timing is everything." *J Clin Endocrinol Metab* 85(2): 804–807.

Hill, K. 1993. "Life history theory and evolutionary anthropology." *Evol Anthropol* 2: 78–88.

Hill, K., C. Boesch et al. 2001. "Mortality rates among wild chimpanzees." *J Hum Evol* 40(5): 437–450.

Hill, K., and A. Hurtado. 1996. *Ache Life History: The Ecology and Demography of a Foraging People*. Hawthorne, NY: Aldine Transaction.

Hindorff, L.A., H.A. Junkins, P.N. Hall et al. "A catalog of published genome-wide association studies." Available at: www.genome.gov/gwastudies. Accessed Nov 1 2010; retrieved Nov 1, 2010.

Hirsch, H.J., T. Eldar-Geva, F. Benarroch, O. Rubinstein, and V. Gross-Tsur. 2009. "Primary testicular dysfunction is a major contributor to abnormal pubertal development in males with Prader-Willi syndrome." *J Clin Endocrinol Metab* 94(7): 2262–2268.

Hochberg, Z. 2002a. "Clinical physiology and pathology of the growth plate." *Best Pract Res Clin Endocrinol Metab* 16(3): 399–419.

Hochberg, Z. 2002b. *Endocrine Control of Bone Maturation*. Basel: Karger.

Hochberg, Z. 2008. "Juvenility in the context of life history theory." *Arch Dis Child* 93(6): 534–539.

Hochberg, Z. 2009. "Evo-devo of child growth II: Human life history and transition between its phases." *Eur J Endocrinol* 160(2): 135–141.

Hochberg, Z. 2010. "Evo-devo of child growth III: Premature juvenility as an evolutionary trade-off." *Horm Res Pediatr* 73: 43–47.

Hochberg, Z., and K. Albertsson-Wikland. 2008. "Evo-devo of infantile and childhood growth." *Pediatr Res* 64(1): 2–7.

Hochberg, Z., M. Aviram et al. 1997. "Decreased sensitivity to insulin-like growth factor I in Turner's syndrome: A study of monocytes and T lymphocytes." *Eur J Clin Invest* 27(7): 543–547.

Hochberg, Z., and A. Etzioni. 1995. "Genetic selection in nonclassical adrenal hyperplasia." *J Clin Endocrinol Metab* 80(1): 325–326.

Hochberg, Z., R. Feil et al. 2011. "Child health, developmental plasticity, and epigenetic programming." *Endocr Rev* 32(2): 159–224.

Hochberg, Z., A. Gawlik, and R.S. Walker. 2011. "Evolutionary fitness as a function of pubertal age in 22 subsistence-based traditional societies." *Int J Pediatr Endocrinol* 2 (online publication).

Hochberg, Z., I. Khaesh-Goldberg et al. 2005. "Differences in infantile growth patterns in Turner syndrome girls with and without spontaneous puberty." *Horm Metab Res* 37(4): 236–241.

Hochberg, Z., R. Perlman et al. 1983. "Insulin regulates placental lactogen and estradiol secretion by cultured human term trophoblast." *J Clin Endocrinol Metab* 57(6): 1311–1313.

Hochberg, Z., R. Feil, M. Constancia et al. 2011. "Child health, developmental plasticity, and epigenetic programming." *Endocr Rev* 32(2): 159–224.

Hockfield, S., and R.D. McKay. 1985. "Identification of major cell classes in the developing mammalian nervous system." *J Neurosci* 5(12): 3310–3328.

Holmberg, J., A. Armulik, K.A. Senti et al. 2005. "Ephrin-A2 reverse signaling negatively regulates neural progenitor proliferation and neurogenesis." *Genes Dev* 19(4): 462–471.

Horvath, T.L., S. Diano et al. 2001. "Minireview: Ghrelin and the regulation of energy balance—a hypothalamic perspective." *Endocrinology* 142(10): 4163–4169.

Howell, N. 2000. *Demography of the Dobe !Kung*. New York: Aldine de Gruyter.

Howell, N. 2010. *Life Histories of the Dobe !Kung*. Berkley: University of California Press.

Huber, K., B. Bruhl, F. Guillemot et al. 2002. "Development of chromaffin cells depends on MASH1 function." *Development* 129(20): 4729–4738.

Huff, D.S., F. Hadziselimovic et al. 1987. "Germ cell counts in semithin sections of biopsies of 115 unilaterally cryptorchid testes. The experience from the Children's Hospital of Philadelphia." *Eur J Pediatr* 146(Suppl 2): S25–27.

Humphrey, L.T. 1998. "Growth patterns in the modern human skeleton." *Am J Phys Anthropol* 105(1): 57–72.

Hunt, P.J., E.M. Gurnell et al. 2000. "Improvement in mood and fatigue after dehydroepiandrosterone replacement in Addison's disease in a randomized, double blind trial." *J Clin Endocrinol Metab* 85(12): 4650–4656.

Ibanez, L., J. Dimartino-Nardi et al. 2000. "Premature adrenarche—normal variant or forerunner of adult disease?" *Endocr Rev* 21(6): 671–696.

Ibanez, L., R. Jimenez et al. 2006. "Early puberty–menarche after precocious pubarche: Relation to prenatal growth." *Pediatrics* 117(1): 117–121.

Ibanez, L., A. Lopez-Bermejo et al. 2008. "Metformin treatment for four years to reduce total and visceral fat in low birth weight girls with precocious pubarche." *J Clin Endocrinol Metab* 93(5): 1841–1845.

Ibancz, L., K. Ong et al. 2001. "Insulin gene variable number of tandem repeat genotype and the low birth weight, precocious pubarche, and hyperinsulinism sequence." *J Clin Endocrinol Metab* 86(12): 5788–5793.

Ibanez, L., K.K. Ong et al. 2003. "Androgen receptor gene CAG repeat polymorphism in the development of ovarian hyperandrogenism." *J Clin Endocrinol Metab* 88(7): 3333–3338.

Ibanez, L., R. Virdis et al. 1992. "Natural history of premature pubarche: An auxological study." *J Clin Endocrinol Metab* 74(2): 254–257.

Inamdar, N.M., K.C. Ehrlich, and M. Ehrlich. 1991. "CpG methylation inhibits binding of several sequence-specific DNA-binding proteins from pea, wheat, soybean and cauliflower." *Plant Mol Biol* 17(1): 111–123.

Ito, S., A.C. D'Alessio, O.V. Taranova et al. 2010. "Role of Tet proteins in 5mC to 5hmC conversion, ES-cell self-renewal and inner cell mass specification." *Nature* 466(7310): 1129–1133.

Jackson, A.A., R.L. Dunn et al. 2002. "Increased systolic blood pressure in rats induced by a maternal low-protein diet is reversed by dietary supplementation with glycine." *Clin Sci (Lond)* 103(6): 633–639.

Jackson, E.L., J.M. Garcia-Verdugo, S. Gil-Perotin et al. 2006. "PDGFR alpha-positive B cells are neural stem cells in the adult SVZ that form glioma-like growths in response to increased PDGF signaling." *Neuron* 51(2): 187–199.

Janson, C.H., and C.P. van Schaik. 1993. "Ecological risk aversion in juvenile primates: Slow and steady wins the race." *Juvenile Primates: Life History, Development, and Behavior*, edited by M.E. Pereira and L.A. Fairbanks. New York: Oxford University Press, pp. 57–74.

Jetz, W., C. Carbone et al. 2004. "The scaling of animal space use." *Science* 306(5694): 266–268.

Jiang, Y.H., J. Bressler et al. 2004. "Epigenetics and human disease." *Annu Rev Genomics Hum Genet* 5: 479–510.

Jiao, J.W., D.A. Feldheim, D.F. Chen et al. 2008. "Ephrins as negative regulators of adult neurogenesis in diverse regions of the central nervous system." *Proc Natl Acad Sci USA* 105(25): 8778–8783.

Jin, K., M. Minami, J.Q. Lan et al. 2001. "Neurogenesis in dentate subgranular zone and rostral subventricular zone after focal cerebral ischemia in the rat." *Proc Natl Acad Sci USA* 98(8): 4710–4715.

Jin, S.G., C. Guo, and G.P. Pfeifer. 2008. "GADD45A does not promote DNA demethylation." *PLoS Genet* 4(3): e1000013.

Jin, S.G., X. Wu, A.X. Li, and G.P. Pfeifer. 2011. "Genomic mapping of 5-hydroxymethylcytosine in the human brain." *Nucleic Acids Res.* doi: 10.1093/nar/gkr120.

Jirtle, R.L., and M.K. Skinner. 2007. "Environmental epigenomics and disease susceptibility." *Nat Rev Genet* 8(4): 253–262.

Johe, K.K., T.G. Hazel et al. 1996. "Single factors direct the differentiation of stem cells from the fetal and adult central nervous system." *Genes Dev* 10(24): 3129–3140.

Jolley, C.D. 2003. "Failure to thrive." *Curr Probl Pediatr Adolesc Health Care* 33(6): 183–206.

Jost, J.P. 1993. "Nuclear extracts of chicken embryos promote an active demethylation of DNA by excision repair of 5-methyldeoxycytidine." *Proc Natl Acad Sci USA* 90(10): 4684–4688.

Juul, A., T. Scheike et al. 2002. "Low serum insulin-like growth factor I is associated with increased risk of ischemic heart disease: A population-based case-control study." *Circulation* 106(8): 939–944.

Kaminsky, Z., S.C. Wang, and A. Petronis. 2006. "Complex disease, gender and epigenetics." *Ann Med* 38(8): 530–544.

Kaminsky, Z.A., T. Tang, and S.C. Wang. 2009. "DNA methylation profiles in monozygotic and dizygotic twins." *Nat Genet* 41(2): 240–245.

Kangaspeska, S., B. Stride, R. Metivier et al. 2008. "Transient cyclical methylation of promoter DNA." *Nature* 452(7183): 112–115.

Kaplan, H.S., and J.B. Lancaster. 2003. "An evolutionary and ecological analysis of human fertility, mating patterns, and parental investment." In: *Offspring: Human Fertility Behavior in Biodemographic Perspective*, edited by K. Wachter and R.A. Bulatao, 170–223. Washington: The National Academies Press.

Kaplan, H.S., and A.J. Robson. 2002. "The emergence of humans: The coevolution of intelligence and longevity with intergenerational transfers." *Proc Natl Acad Sci USA* 99(15): 10221–10226.

Kaplan, M.S., and D.H. Bell. 1984. "Mitotic neuroblasts in the 9-day-old and 11-month-old rodent hippocampus." *J Neurosci* 4(6): 1429–1441.

Kaplowitz, P.B., E.J. Slora et al. 2001. "Earlier onset of puberty in girls: Relation to increased body mass index and race." *Pediatrics* 108(2): 347–353.

Kappeler, L., C. De Magalhaes Filho et al. 2009. "Early postnatal nutrition determines somatotropic function in mice." *Endocrinology* 150(1): 314–323.

Kaprio, J., A. Rimpela, T. Winter et al. 1995. "Common genetic influences on BMI and age at menarche." *Hum Biol* 67(5): 739–753.

Karlberg, J. 1987. "On the modelling of human growth." *Stat Med* 6(2): 185–192.

Karlberg, J. 1989. "A biologically-oriented mathematical model (ICP) for human growth." *Acta Paediatr Scand Suppl* 350: 70–94.

Karlberg, J., and K. Albertsson-Wikland. 1988. "Infancy growth pattern related to growth hormone deficiency." *Acta Paediatr Scand* 77(3): 385–391.

Karlberg, J., K. Albertsson-Wikland et al. 1991. "Growth in infancy and childhood in girls with Turner's syndrome." *Acta Paediatr Scand* 80(12): 1158–1165.

Karlberg, J., I. Engstrom et al. 1987. "Analysis of linear growth using a mathematical model. I. From birth to three years." *Acta Paediatr Scand* 76(3): 478–488.

Karlberg, J., F. Jalil et al. 1994. "Linear growth retardation in relation to the three phases of growth." *Eur J Clin Nutr* 48(Suppl 1): S25–43; discussion S43–44.

Karsenty, G. 2006. "Convergence between bone and energy homeostases: Leptin regulation of bone mass." *Cell Metab* 4(5): 341–348.

Kemkes-Grottenthaler, A. 2005. "The short die young: The interrelationship between stature and longevity-evidence from skeletal remains." *Am J Phys Anthropol* 128(2): 340–347.

Kempermann, G., H.G. Kuhn, and F.H. Gage. 1997. "More hippocampal neurons in adult mice living in an enriched environment." *Nature* 386(6624): 493–495.

Kempermann, G., H.G. Kuhn, and F.H. Gage. 1998. "Experience-induced neurogenesis in the senescent dentate gyrus." *J Neurosci* 18(9): 3206–3212.

Kennedy, G.E. 2005. "From the ape's dilemma to the weanling's dilemma: Early weaning and its evolutionary context." *J Hum Evol* 48(2): 123–145.

Kerem, N., H. Guttmann et al. 2001. "The autosomal dominant trait of obesity, acanthosis nigricans, hypertension, ischemic heart disease and diabetes type 2." *Horm Res* 55(6): 298–304.

Key, C., and L.C. Aiello. 2000. "A prisoner's dilemma model of the evolution of paternal care." *Folia Primatol* 71: 77–92.

Kimura, M. 1968. "Evolutionary rate at the molecular level." *Nature* 217: 624–626.

Kirkwood, T.B.L., P. Kapahi et al. 2000. "Evolution, stress, and longevity." *J Anat* 197 Pt 4: 587–590.

Kirkwood, T.B.L., and M.R. Rose. 1991. "Evolution of senescence: Late survival sacrificed for reproduction." *Philos Trans R Soc Lond B Biol Sci* 332(1262): 15–24.

Kleiber, M. 1932. "Body size and metabolism." *Hilgardia* 6: 315–353.

Klein, K.O., J. Baron et al. 1994. "Estrogen levels in childhood determined by an ultrasensitive recombinant cell bioassay." *J Clin Invest* 94(6): 2475–2480.

Kobata, R., H. Tsukahara et al. 2008. "High levels of growth factors in human breast milk." *Early Hum Dev* 84(1): 67–69.

Korkeila, M., J. Kaprio, A. Rissanen, and M. Koskenvuo. 1991. "Effects of gender and age on the heritability of body mass index." *Int J Obes* 15(10): 647–654.

Korth-Schutz, S., L.S. Levine et al. 1976. "Dehydroepiandrosterone sulfate (DS) levels, a rapid test for abnormal adrenal androgen secretion." *J Clin Endocrinol Metab* 42(6): 1005–1013.

Kriaucionis, S., and N. Heintz. 2009. "The nuclear DNA base 5-hydroxymethylcytosine is present in Purkinje neurons and the brain." *Science* 324(5929): 929–930.

Kristrom, B., Hochberg, Z., and K. Albertsson-Wikland. 2007. "Delayed infancy-childhood spurt (DICS) in idiopathic short stature (ISS)." *Horm Res* 68(Suppl 1): 168.

Kroboth, P.D., F.S. Salek et al. 1999. "DHEA and DHEA-S: A review." *J Clin Pharmacol* 39(4): 327–348.

Kucharski, R., J. Maleszka, S. Foret, and R. Maleszka. 2008. "Nutritional control of reproductive status in honeybees via DNA methylation." *Science* 319(5871): 1827–1830.

Kuhn, H.G., H. Dickinson-Anson, and F.H. Gage. 1996. "Neurogenesis in the dentate gyrus of the adult rat: Age-related decrease of neuronal progenitor proliferation." *J Neurosci* 16(6): 2027–2033.

Kuzawa, C.W. 1998. "Adipose tissue in human infancy and childhood: An evolutionary perspective." *Am J Phys Anthropol* Suppl 27: 177–209.

Kuzawa, C.W. 2007. "Developmental origins of life history: Growth, productivity, and reproduction." *Am J Hum Biol* 19(5): 654–661.

Lahav, R., C. Ziller, E. Dupin, and N.M. Le Douarin. 1996. "Endothelin 3 promotes neural crest cell proliferation and mediates a vast increase in melanocyte number in culture." *Proc Natl Acad Sci USA* 93(9): 3892–3897.

Lahdenpera, M., V. Lummaa et al. 2004. "Fitness benefits of prolonged post-reproductive lifespan in women." *Nature* 428(6979): 178–181.

Lakshman, R., N.G. Forouhi, S.J. Sharp et al. 2009. "Early age at menarche associated with cardiovascular disease and mortality." *J Clin Endocrinol Metab* 94: 4953–4960.

Lancaster, J.B., and C.S. Lancaster. 1983. "Parental investment: The hominid adaptation." *How Humans Adapt: A Biocultural Odyssey*, edited by D.J. Ortner, 33–65. Washington, DC: Smithsonian Institution Press.

Lantz, H., L.E. Bratteby et al. 2008. "Body composition in a cohort of Swedish adolescents aged 15, 17 and 20.5 years." *Acta Paediatr* 97(12): 1691–1697.

Lappalainen, S., P. Utriainen et al. 2008. "Androgen receptor gene CAG repeat polymorphism and X-chromosome inactivation in children with premature adrenarche." *J Clin Endocrinol Metab* 93(4): 1304–1309.

Lappalainen, S., R. Voutilainen et al. 2009. "Genetic variation of FTO and TCF7L2 in premature adrenarche." *Metabolism* 58(9): 1263–1269.

Lasker, G.W. 1969. "Human biological adaptability. The ecological approach in physical anthropology." *Science* 166(3912): 1480–1486.

Latta, L.C. IV, J.W. Bakelar, R.A. Knapp, and M.E. Pfrender. 2007. "Rapid evolution in response to introduced predators II: The contribution of adaptive plasticity." *BMC Evol Biol* 7: 21.

Lebel, C., L. Walker et al. 2008. "Microstructural maturation of the human brain from childhood to adulthood." *Neuroimage* 40(3): 1044–1055.

Lee, A., J.D. Kessler, T.-A. Read et al. 2005. "Isolation of neural stem cells from the postnatal cerebellum." *Nat Neurosci* 8(6): 723–729.

Lee, C.J., G.S. Lawler et al. 1987. "Nutritional status of middle-aged and elderly females in Kentucky in two seasons: Part 2. Hematological parameters." *J Am Coll Nutr* 6(3): 217–222.

Lee, P.C., P. Majluf, I.J. Gordon. 1991. "Growth, weaning and maternal investment from a comparative perspective." *Journal of Zoology* 225: 99–114.

Lee, R. 1976. *The !Kung San*. Cambridge: Cambridge University Press.

Lee, R., and I. DeVore, eds. 1976. *Kalahari Hunter-Gatherers: Studies of the !Kung San and Their Neighbors*. Cambridge: Harvard University Press.

Leger, J., J.F. Oury et al. 1996. "Growth factors and intrauterine growth retardation. I. Serum growth hormone, insulin-like growth factor (IGF)-I, IGF-II, and IGF binding protein 3 levels in normally grown and growth-retarded human fetuses during the second half of gestation." *Pediatr Res* 40(1): 94–100.

Leigh, S.R., and B.T. Shea. 1996. "Ontogeny of body size variation in African apes." *Am J Phys Anthropol* 99(1): 43–65.

Lendahl, U., L.B. Zimmerman, and R.D. McKay. 1990. "CNS stem cells express a new class of intermediate filament protein." *Cell* 60(4): 585–595.

Leonard, W.R., and M.L. Robertson. 1997. "Comparative primate energetics and hominid evolution." *Am J Phys Anthropol* 102(2): 265–281.

Leonard, W.R., J.J. Snodgrass, and M.L. Robertson. 2007. "Effects of brain evolution on human nutrition and metabolism." *Annu Rev Nutr* 27: 311–327.

Leutenegger, W. 1973. "Maternal-fetal weight relationships in primates." *Folia Primatol (Basel)* 20(4): 280–293.

Levenson, J.M., T.L. Roth, F.D. Lubin, and C.A. Miller. 2006. "Evidence that DNA (cytosine-5) methyltransferase regulates synaptic plasticity in the hippocampus." *J Biol Chem* 281(23): 15763–15773.

Levison, S.W., and J.E. Goldman. 1993. "Both oligodendrocytes and astrocytes develop from progenitors in the subventricular zone of postnatal rat forebrain." *Neuron* 10(2): 201–212.

Li, Z., W. Xiao et al. 2002. "Identification of a protein essential for a major pathway used by human cells to avoid UV-induced DNA damage." *Proc Natl Acad Sci USA* 99(7): 4459–4464.

Licinio, J., S. Caglayan, and M. Ozata. 2004. "Phenotypic effects of leptin replacement on morbid obesity, diabetes mellitus, hypogonadism, and behavior in leptin-deficient adults." *Proc Natl Acad Sci USA* 101(13): 4531–4536.

Lillycrop, K.A., E.S. Phillips et al. 2005. "Dietary protein restriction of pregnant rats induces and folic acid supplementation prevents epigenetic modification of hepatic gene expression in the offspring." *J Nutr* 135(6): 1382–1386.

Lillycrop, K.A., E.S. Phillips et al. 2008. "Feeding pregnant rats a protein-restricted diet persistently alters the methylation of specific cytosines in the hepatic PPAR alpha promoter of the offspring." *Br J Nutr* 100(2): 278–282.

Lillycrop, K.A., J.L. Slater-Jefferies et al. 2007. "Induction of altered epigenetic regulation of the hepatic glucocorticoid receptor in the offspring of rats fed a protein-restricted diet during pregnancy suggests that reduced DNA methyltransferase-1 expression is involved in impaired DNA methylation and changes in histone modifications." *Br J Nutr* 97(6): 1064–1073.

Lindgren, A.C., B. Barkeling et al. 2000. "Eating behavior in Prader-Willi syndrome, normal weight, and obese control groups." *J Pediatr* 137(1): 50–55.

Linthorst, A.C., C. Flachskamm, P. Muller-Preuss et al. 1995. "Effect of bacterial endotoxin and interleukin-1 beta on hippocampal serotonergic neurotransmission, behavioral activity, and free corticosterone levels: An *in vivo* microdialysis study." *J Neurosci* 15(4): 2920–2934.

Lister, R., M. Pelizzola, R.D. Hawkins et al. 2009. "Human DNA methylomes at base resolution show widespread epigenomic differences." *Nature* 462(7271): 315–322.

Little, M.T., and P. Hahn. 1990. "Diet and metabolic development." *FASEB J* 4(9): 2605–2611.

Liu, D., J. Diorio, B. Tannenbaum et al. 1997. "Maternal care, hippocampal glucocorticoid receptors, and hypothalamic-pituitary-adrenal responses to stress." *Science* 277(5332): 1659–1662.

Liu, J., X. Qiao et al. 2004. "Motilin in human milk and its elevated plasma concentration in lactating women." *J Gastroenterol Hepatol* 19(10): 1187–1191.

Liu, J., K. Solway, and F.R. Sharp. 1998. "Increased neurogenesis in the dentate gyrus after transient global ischemia in gerbils." *J Neurosci* 18(19): 7768–7778.

Liu, X., J.P. Chism et al. 1999. "Correlation of biliary excretion in sandwich-cultured rat hepatocytes and *in vivo* in rats." *Drug Metab Dispos* 27(6): 637–644.

Liu, X., L. Serova, R. Kvetnansky, and E.L. Sabban. 2008. "Identifying the stress transcriptome in the adrenal medulla following acute and repeated immobilization." *Ann NY Acad Sci* 1148: 1–28.

Liu, Y., K. Albertsson-Wikland et al. 2000. "Long-term consequences of early linear growth retardation (stunting) in Swedish children." *Pediatr Res* 47(4 Pt 1): 475–480.

Liu, Y.X., F. Jalil et al. 1998. "Growth stunting in early life in relation to the onset of the childhood component of growth." *J Pediatr Endocrinol Metab* 11(2): 247–260.

Lo, L.C., S.J. Birren, and D.J. Anderson. 1991. "V-myc immortalization of early rat neural crest cells yields a clonal cell line which generates both glial and adrenergic progenitor cells." *Dev Biol* 145(1): 139–153.

Locke, J.L., and B. Bogin. 2006. "Language and life history: A new perspective on the development and evolution of human language." *Behav Brain Sci* 29(3): 259–280; discussion 280–325.

Lois, C., and A. Alvarez-Buylla. 1993. "Proliferating subventricular zone cells in the adult mammalian forebrain can differentiate into neurons and glia." *Proc Natl Acad Sci USA* 90(5): 2074–2077.

Lopez-Maury, L., S. Marguerat et al. 2008. "Tuning gene expression to changing environments: From rapid responses to evolutionary adaptation." *Nat Rev Genet* 9(8): 583–593.

Loria, R.M., and D.A. Padgett. 1992. "Mobilization of cutaneous immunity for systemic protection against infections." *Ann NY Acad Sci* 650: 363–366.

Louis, J., C. Cannard et al. 1997. "Sleep ontogenesis revisited: A longitudinal 24-hour home polygraphic study on 15 normal infants during the first two years of life." *Sleep* 20(5): 323–333.

Lovejoy, C. 1981. "The origin of man." *Science* 211: 341–350.

Lucarelli, M., A. Fuso, R. Strom, and S. Scarpa. 2001. "The dynamics of myogenin site-specific demethylation is strongly correlated with its expression and with muscle differentiation." *J Biol Chem* 276(10): 7500–7506.

Lunn, P.G. 1992. "Breast-feeding patterns, maternal milk output and lactational infecundity." *J Biosoc Sci* 24(3): 317–324.

Luskin, M.B. 1993. "Restricted proliferation and migration of postnatally generated neurons derived from the forebrain subventricular zone." *Neuron* 11(1): 173–189.

Lyko, F., S. Foret, R. Kucharski et al. 2010. "The honey bee epigenomes: Differential methylation of brain DNA in queens and workers." *PLoS Biol* 8(11): e1000506.

Ma, D.K., M.H. Jang, J.U. Guo et al. 2009. "Neuronal activity-induced Gadd45b promotes epigenetic DNA demethylation and adult neurogenesis." *Science* 323(5917): 1074–1077.

Magavi, S.S., B.R. Leavitt, and J.D. Macklis. 2000. "Induction of neurogenesis in the neocortex of adult mice." *Nature* 405(6789): 951–955.

Maleszka, R. 2008. "Epigenetic integration of environmental and genomic signals in honey bees: The critical interplay of nutritional, brain and reproductive networks." *Epigenetics* 3(4): 188–192.

Mann, F.D. 1998. "Animal fat and cholesterol may have helped primitive man evolve a large brain." *Perspectives in Biology and Medicine* 41: 417–425.

Manolio, T.A., F.S. Collins, N.J. Cox et al. 2009. "Finding the missing heritability of complex diseases." *Nature* 461(7265): 747–753.

Marean, C. 2010. "Pinnacle Point Cave 13B (Western Cape Province, South Africa) in context: The Cape Floral kingdom, shellfish, and modern human origins." *Journal of Human Evolution* 59: 425–443.

Marmur, R., P.C. Mabie, S. Gokhan et al. 1998. "Isolation and developmental characterization of cerebral cortical multipotent progenitors." *Dev Biol* 204(2): 577–591.

Martens, D.J., R.M. Seaberg, and D. van der Kooy et al. 2002. "*In vivo* infusions of exogenous growth factors into the fourth ventricle of the adult mouse brain increase the proliferation of neural progenitors around the fourth ventricle and the central canal of the spinal cord." *Eur J Neurosci* 16(6): 1045–1057.

Martin, D.S., D.Z. Levett, M.P. Grocott, and H.E. Montgomery. 2010. "Variation in human performance in the hypoxic mountain environment." *Exp Physiol* 95(3): 463–470.

Martin-Gronert, M.S., and S.E. Ozanne. 2010. "Mechanisms linking suboptimal early nutrition and increased risk of type 2 diabetes and obesity." *Journal of Nutrition* 140: 662–666.

Mastronardi, F.G., A. Noor, D.D. Wood et al. 2007. "Peptidyl argininedeiminase 2 CpG island in multiple sclerosis white matter is hypomethylated." *J Neurosci Res* 85(9): 2006–2016.

Matchock, R.L., and E.J. Susman. 2006. "Family composition and menarcheal age: Anti-inbreeding strategies." *Am J Hum Biol* 18(4): 481–491.

McDade, T.W., V. Reyes-Garcia et al. 2008. "Maintenance versus growth: Investigating the costs of immune activation among children in lowland Bolivia." *Am J Phys Anthropol* 136(4): 478–484.

McGowan, P.O., A. Sasaki, A.C. D'Alessio et al. 2009. "Epigenetic regulation of the glucocorticoid receptor in human brain associates with childhood abuse." *Nat Neurosci* 12(3): 342–348.

McGowan, P.O., A. Sasaki, T.C. Huang et al. 2008. "Promoter-wide hypermethylation of the ribosomal RNA gene promoter in the suicide brain." *PLoS ONE* 3(5): e2085.

McGowan, P.O., M. Suderman, A. Sasaki et al. 2011. "Broad epigenetic signature of maternal care in the brain of adult rats." *PLoS One* 6(2): e14739.

McKusick, V. 1956. *Heritable Disorders of Connective Tissue*. St. Louis, MO: Mosby.

McMillen, I.C., B.S. Muhlhausler et al. 2004. "Prenatal programming of postnatal obesity: Fetal nutrition and the regulation of leptin synthesis and secretion before birth." *Proc Nutr Soc* 63(3): 405–412.

Mead, M. 2001. *Coming of Age in Samoa*. New York: Perrenial Classics.

Melzner, I., V. Scott et al. 2002. "Leptin gene expression in human preadipocytes is switched on by maturation-induced demethylation of distinct CpGs in its proximal promoter." *J Biol Chem* 277(47): 45420–45427.

Meyer, D.L., M.S. Kerley et al. 2005. "Growth rate, body composition, and meat tenderness in early vs. traditionally weaned beef calves." *J Anim Sci* 83(12): 2752–2761.

Migliano, A.B., L. Vinicius et al. 2007. "Life history trade-offs explain the evolution of human pygmies." *Proc Natl Acad Sci USA* 104(51): 20216–20219.

Mill, J., T. Tang, Z. Kaminsky et al. 2008. "Epigenomic profiling reveals DNA-methylation changes associated with major psychosis." *Am J Hum Genet* 82(3): 696–711.

Miller, C.A., and J.D. Sweatt. 2007. "Covalent modification of DNA regulates memory formation." *Neuron* 53(6): 857–869.

Miller, W.L. 1999. "The molecular basis of premature adrenarche: An hypothesis." *Acta Paediatr Suppl* 88(433): 60–66.

Milsom, S.R., W.F. Blum et al. 2008. "Temporal changes in insulin-like growth factors I and II and in insulin-like growth factor binding proteins 1, 2, and 3 in human milk." *Horm Res* 69(5): 307–311.

Mitamura, R., K. Yano et al. 1999. "Diurnal rhythms of luteinizing hormone, follicle-stimulating hormone, and testosterone secretion before the onset of male puberty." *J Clin Endocrinol Metab* 84(1): 29–37.

Mitamura, R., K. Yano et al. 2000. "Diurnal rhythms of luteinizing hormone, follicle-stimulating hormone, testosterone, and estradiol secretion before the onset of female puberty in short children." *J Clin Endocrinol Metab* 85(3): 1074–1080.

Moldovan, N.I. 2005. "Functional adaptation: The key to plasticity of cardiovascular 'stem' cells?" *Stem Cells Dev* 14(2): 111–121.

Molinari, L., R.H. Largo et al. 1980. "Analysis of the growth spurt at age seven (mid-growth spurt)." *Helv Paediatr Acta* 35(4): 325–334.

Monje, M.L., S. Mizumatsu, J.R. Fike, and T.D. Palmer. 2002. "Irradiation induces neural precursor-cell dysfunction." *Nat Med* 8(9): 955–962.

Montagne, L., G. Boudry et al. 2007. "Main intestinal markers associated with the changes in gut architecture and function in piglets after weaning." *Br J Nutr* 97(1): 45–57.

Montemitro, E., P. Franco et al. 2008. "Maturation of spontaneous arousals in healthy infants." *Sleep* 31(1): 47–54.

Morales, A.J., J.J. Nolan et al. 1994. "Effects of replacement dose of dehydroepiandrosterone in men and women of advancing age." *J Clin Endocrinol Metab* 78(6): 1360–1367.

Moritz, K.M., W.M. Boon et al. 2005. "Glucocorticoid programming of adult disease." *Cell Tissue Res* 322(1): 81–88.

Moss, E.G., R.C. Lee, and V. Ambros. 1997. "The cold shock domain protein LIN-28 controls developmental timing in C. elegans and is regulated by the lin-4 RNA." *Cell* 88(5): 637–646.

Moyes, C.D. 1976. "Adverse factors affecting growth of schoolchildren in St. Helena." *Arch Dis Child* 51(6): 435–438.

Moyes, C.D. 1981. "Stature and birth rank. A study of schoolchildren in St. Helena." *Arch Dis Child* 56(2): 116–120.

Muller, G.B. 2007. "Evo-devo: Extending the evolutionary synthesis." *Nat Rev Genet* 8(12): 943–949.

Muller, J., and N.E. Skakkebaek. 1983. "Quantification of germ cells and seminiferous tubules by stereological examination of testicles from 50 boys who suffered from sudden death." *Int J Androl* 6(2): 143–156.

Murakami, M., K. Kawai et al. 1988. "[Correlation between breast development and hormone profiles in puberal girls]." *Nippon Sanka Fujinka Gakkai Zasshi* 40(5): 561–567.

Murgatroyd, C., A.V. Patchev, Y. Wu et al. 2009. "Dynamic DNA methylation programs persistent adverse effects of early-life stress." *Nat Neurosci* 12(12): 1559–1566.

Myatt, L. 2006. "Placental adaptive responses and fetal programming." *J Physiol* 572(Pt 1): 25–30.

Nakatomi, H., T. Kuriu, S. Okabe et al. 2002. "Regeneration of hippocampal pyramidal neurons after ischemic brain injury by recruitment of endogenous neural progenitors." *Cell* 110(4): 429–441.

Nan, X., F.J. Campoy, and A. Bird. 1997. "MeCP2 is a transcriptional repressor with abundant binding sites in genomic chromatin." *Cell* 88(4): 471–481.

Neel, J.V. 1962. "Diabetes mellitus: A 'thrifty' genotype rendered detrimental by 'progress'?" *American Journal of Human Genetics* 14: 353–362.

Ng, H.H., Y. Zhang, B. Henrich et al. 1999. "MBD2 is a transcriptional repressor belonging to the MeCP1 histone deacetylase complex" [see comments]. *Nat Genet* 23(1): 58–61.

Ng, S.F., R.C. Lin et al. 2010. "Chronic high-fat diet in fathers programs beta-cell dysfunction in female rat offspring." *Nature* 467(7318): 963–966.

Nijhout, H.F. 2003. "Development and evolution of adaptive polyphenisms." *Evol Dev* 5(1): 9–18.

Niklasson, A., and K. Albertsson-Wikland. 2008. "Continuous growth reference from 24th week of gestation to 24 months by gender." *BMC Pediatr* 8: 8.

Noonan, J.A., R. Raaijmakers et al. 2003. "Adult height in Noonan syndrome." *Am J Med Genet A* 123(1): 68–71.

Norjavaara, E., C. Ankarberg et al. 1996. "Diurnal rhythm of 17 beta-estradiol secretion throughout pubertal development in healthy girls: Evaluation by a sensitive radioimmunoassay." *J Clin Endocrinol Metab* 81(11): 4095–4102.

Oberlander, T.F., J. Weinberg, M. Papsdorf et al. 2008. "Prenatal exposure to maternal depression, neonatal methylation of human glucocorticoid receptor gene (NR3C1) and infant cortisol stress responses." *Epigenetics* 3(2): 97–106.

Ogueh, O., S. Sooranna et al. 2000. "The relationship between leptin concentration and bone metabolism in the human fetus." *J Clin Endocrinol Metab* 85(5): 1997–1999.

Ohta, T. 1976. "Role of very slightly deleterious mutations in molecular evolution and polymorphism." *Theoretical Population Biology* 10: 254–275.

Okano, M., S. Xie, and E. Li. 1998. "Cloning and characterization of a family of novel mammalian DNA (cytosine-5) methyltransferases" [letter]. *Nat Genet* 19(3): 219–220.

Ong, K.K., and R.J. Loos. 2006. "Rapid infancy weight gain and subsequent obesity: Systematic reviews and hopeful suggestions." *Acta Paediatr* 95(8): 904–908.

Ong, K.K., N. Potau et al. 2004. "Opposing influences of prenatal and postnatal weight gain on adrenarche in normal boys and girls." *J Clin Endocrinol Metab* 89(6): 2647–2651.

Oswald, J., S. Engemann, N. Lane et al. 2000. "Active demethylation of the paternal genome in the mouse zygote." *Curr Biol* 10(8): 475–478.

Ourednik, J., V. Ourednik, W.P. Lynch et al. 2002. "Neural stem cells display an inherent mechanism for rescuing dysfunctional neurons." *Nat Biotechnol* 20(11): 1103–1110.

Palmer, T.D., J. Ray, and F.H. Gage. 1995. "FGF-2-responsive neuronal progenitors reside in proliferative and quiescent regions of the adult rodent brain." *Mol Cell Neurosci* 6(5): 474–486.

Palmer, T.D., J. Takahashi, and F.H. Gag. 1997. "The adult rat hippocampus contains primordial neural stem cells." *Mol Cell Neurosci* 8(6): 389–404.

Palmert, M.R., D.L. Hayden et al. 2001. "The longitudinal study of adrenal maturation during gonadal suppression: Evidence that adrenarche is a gradual process." *J Clin Endocrinol Metab* 86(9): 4536–4542.

Panici, J.A., J.M. Harper et al. 2010. "Early life growth hormone treatment shortens longevity and decreases cellular stress resistance in long-lived mutant mice." *FASEB J* 24(12): 5073–5079.

Papanicolaou, D.A., R.L. Wilder, S.C. Manolagas, and G.P. Chrousos. 1998. "The pathophysiologic roles of interleukin-6 in humans." *Ann Intern Med* 128: 127–137.

Parent, A.S., G. Teilmann, C. Gong et al. 2003. "The timing of normal puberty and the age limits of sexual precocity: Variations around the world, secular trends, and changes after migration." *Endocr Rev* 24(5): 668–693.

Parent, J.M., Z.S. Vexler et al. 2002. "Rat forebrain neurogenesis and striatal neuron replacement after focal stroke." *Ann Neurol* 52(6): 802–813.

Parsons, T.J., C. Power, S. Logan, and C.D. Summerbell. 1999. "Childhood predictors of adult obesity: A systematic review." *Int J Obes Relat Metab Disord* 23(Suppl 8): S1–107.

Pasquet, P., A.M. Biyong et al. 1999. "Age at menarche and urbanization in Cameroon: Current status and secular trends." *Ann Hum Biol* 26(1): 89–97.

Paus, T. 2005. "Mapping brain maturation and cognitive development during adolescence." *Trends Cogn Sci* 9(2): 60–68.

Pedersen, C.A. 2004. "Biological aspects of social bonding and the roots of human violence." *Ann NY Acad Sci* 1036: 106–127.

Pelizzola, M., Y. Koga, A.E. Urban et al. 2008. "MEDME: An experimental and analytical methodology for the estimation of DNA methylation levels based on microarray derived MeDIP-enrichment." *Genome Res* 18(10): 1652–1659.

Pembrey, M.E., L.O. Bygren et al. 2006. "Sex-specific, male-line transgenerational responses in humans." *Eur J Hum Genet* 14(2): 159–166.

Pere, A., J. Perheentupa et al. 1995. "Follow up of growth and steroids in premature adrenarche." *Eur J Pediatr* 154(5): 346–352.

Petronis, A. 2006. "Epigenetics and twins: Three variations on the theme." *Trends Genet* 22(7): 347–350.

Petry, C.J., K.K. Ong et al. 2005. "Association of aromatase (CYP 19) gene variation with features of hyperandrogenism in two populations of young women." *Hum Reprod* 20(7): 1837–1843.

Pfeiffer, S., and J. Sealy. 2006. "Body size among Holocene foragers of the Cape Ecozone, southern Africa." *Am J Phys Anthropol* 129(1): 1–11.

Piaget, J. 1936. *The Origins of Intelligence in Children*. New York: International University Press.

Piaget, J. 1952. *The Origins of Intelligence in Children*. New York: International University Press.

Popp, C., W. Dean, S. Feng et al. 2010. "Genome-wide erasure of DNA methylation in mouse primordial germ cells is affected by AID deficiency." *Nature* 463(7284): 1101–1105.

Poulter, M.O., L. Du, I.C. Weaver et al. 2008. "GABAA receptor promoter hypermethylation in suicide brain: Implications for the involvement of epigenetic processes." *Biol Psychiatry* 64(8): 645–652.

Power, C., C. Hertzman, C. Matthews et al. 1997. "Social differences in health: Life-cycle effects between ages 23 and 33 in the 1958 British birth cohort." *Am J Public Health* 87(9): 1499–1503.

Power, C., B.J. Jefferis, O. Manor et al. 2006. "The influence of birth weight and socioeconomic position on cognitive development: Does the early home and learning environment modify their effects?" *J Pediatr* 148(1): 54–61.

Power, C., L. Li, and C. Hertzman. 2006. "Associations of early growth and adult adiposity with patterns of salivary cortisol in adulthood." *J Clin Endocrinol Metab* 91(11): 4264–4270.

Powers, J.F., M.J. Evinger, J. Zhi et al. 2007. "Pheochromocytomas in Nf1 knockout mice express a neural progenitor gene expression profile." *Neuroscience* 147(4): 928–937.

Pratt, J.H., A.K. Manatunga et al. 1994. "Familial influences on the adrenal androgen excretion rate during the adrenarche." *Metabolism* 43(2): 186–189.

Prentice, A.M., B.J. Hennig, and A.J. Fulford. 2008. "Evolutionary origins of the obesity epidemic: natural selection of thrifty genes or genetic drift following predation release?" *Int J Obes (Lond)* 32: 1607–1610.

Price, T.D., A. Qvarnstrom et al. 2003. "The role of phenotypic plasticity in driving genetic evolution." *Proc Biol Sci* 270(1523): 1433–1440.

Privat, A., and M.J. Drian. 1975. "Specificity of the formation of the mossy fibre-granule cell synapse in the rat cerebellum. An *in vitro* study." *Brain Res* 88(3): 518–524.

Prokopenko, I., M.I. McCarthy, and C.M. Lindgren. 2008. "Type 2 diabetes: New genes, new understanding." *Trends in Genetics* 24: 613–621.

Qamra, S.R., S. Mehta et al. 1991. "A study of relation between physical growth and sexual maturity in girls. V." *Indian Pediatr* 28(3): 265–272.

Ragusa, L., C. Romano et al. 1992. "Growth hormone subnormality in Down syndrome." *Am J Med Genet* 43(5): 894–895.

Rai, K., S. Chidester, C.V. Zavala et al. 2007. "Dnmt2 functions in the cytoplasm to promote liver, brain, and retina development in zebrafish." *Genes Dev* 21(3): 261–266.

Rai, K., I.J. Huggins, S.R. James et al. 2008. "DNA demethylation in zebrafish involves the coupling of a deaminase, a glycosylase, and gadd45." *Cell* 135(7): 1201–1212.

Raia, P., and S. Meiri. 2006. "The island rule in large mammals: Paleontology meets ecology." *Evolution Int J Org Evolution* 60(8): 1731–1742.

Ramchandani, S., S.K. Bhattacharya, N. Cervoni, and M. Szyf. 1999. "DNA methylation is a reversible biological signal." *Proc Natl Acad Sci USA* 96(11): 6107–6112.

Ranke, M.B., P. Heidemann et al. 1988. "Noonan syndrome: Growth and clinical manifestations in 144 cases." *Eur J Pediatr* 148(3): 220–227.

Rauch, T.A., X. Wu, X. Zhong et al. 2009. "A human B cell methylome at 100-base pair resolution." *Proc Natl Acad Sci USA* 106(3): 671–678.

Ravelli, G.P., Z.A. Stein, and M.W. Susser. 1976. "Obesity in young men after famine exposure in utero and early infancy." *N Engl J Med* 295(7): 349–353.

Razin, A., and H. Cedar. 1977. "Distribution of 5-methylcytosine in chromatin." *Proc Natl Acad Sci USA* 74(7): 2725–2728.

Razin, A., and A.D. Riggs. 1980. "DNA methylation and gene function." *Science* 210(4470): 604–610.

Razin, A., and M. Szyf. 1984. "DNA methylation patterns. Formation and function." *Biochim Biophys Acta* 782(4): 331–342.

Razin, A., M. Szyf, T. Kafri et al. 1986. "Replacement of 5-methylcytosine by cytosine: A possible mechanism for transient DNA demethylation during differentiation." *Proc Natl Acad Sci USA* 83(9): 2827–2831.

Reik, W., M. Constancia et al. 2003. "Regulation of supply and demand for maternal nutrients in mammals by imprinted genes." *J Physiol* 547(Pt 1): 35–44.

Reik, W., W. Dean et al. 2001. "Epigenetic reprogramming in mammalian development." *Science* 293(5532): 1089–1093.

Reiman, E.M., R.J. Caselli, K. Chen et al. 2001. "Declining brain activity in cognitively normal apolipoprotein E varepsilon 4 heterozygotes: A foundation for using positron emission tomography to efficiently test treatments to prevent Alzheimer's disease." *Proc Natl Acad Sci USA* 98: 3334–3339.

Remer, T., K.R. Boye et al. 2003. "Adrenarche and bone modeling and remodeling at the proximal radius: Weak androgens make stronger cortical bone in healthy children." *J Bone Miner Res* 18(8): 1539–1546.

Remer, T., and F. Manz. 2001. "The midgrowth spurt in healthy children is not caused by adrenarche." *J Clin Endocrinol Metab* 86(9): 4183–4186.

Reya, T., S.J. Morrison, M.F. Clarke, and I.L. Weissman. 2001. "Stem cells, cancer, and cancer stem cells." *Nature* 414(6859): 105–111.

Reynolds, B.A., and S. Weiss. 1992. "Generation of neurons and astrocytes from isolated cells of the adult mammalian central nervous system." *Science* 255(5052): 1707–1710.

Richardson, K., and S. Norgate. 2005. "The equal environments assumption of classical twin studies may not hold." *Br J Educ Psychol* 75(Pt 3): 339–350.

Rodda, C., D.A. Jones et al. 1987. "Muscle strength in girls with congenital adrenal hyperplasia." *Acta Paediatr Scand* 76(3): 495–499.

Roff, D.A. 2007. "Contributions of genomics to life-history theory." *Nat Rev Genet* 8(2): 116–125.

Roldan, M.B., C. White et al. 2007. "Association of the GAA1013→GAG polymorphism of the insulin-like growth factor-1 receptor (IGF1R) gene with premature pubarche." *Fertil Steril* 88(2): 410–417.

Rolland-Cachera, M.F., M. Deheeger et al. 2006. "Early adiposity rebound: Causes and consequences for obesity in children and adults." *Int J Obes (Lond)* 30(Suppl 4): S11–17.

Rona, R.J., and S. Chinn. 1982. "National study of health and growth: Social and family factors and obesity in primary schoolchildren." *Ann Hum Biol* 9(2): 131–145.

Rosenfeld, R.G. 2006. "Molecular mechanisms of IGF-I deficiency." *Horm Res* 65(Suppl 1): 15–20.

Rossignol, S., I. Netchine et al. 2008. "Epigenetics in Silver-Russell syndrome." *Best Pract Res Clin Endocrinol Metab* 22(3): 403–414.

Roth, T.L., F.D. Lubin, A.J. Funk, and J.D. Sweatt. 2009. "Lasting epigenetic influence of early-life adversity on the BDNF gene." *Biol Psychiatry* 65(9): 760–769.

Rudolf, M.C., and S. Logan. 2005. "What is the long term outcome for children who fail to thrive? A systematic review." *Arch Dis Child* 90(9): 925–931.

Ruppenthal, G.C., G.L. Arling, H.F. Harlow et al. 1976. "A 10-year perspective of motherless-mother monkey behavior." *J Abnorm Psychol* 85(4): 341–349.

Saleemi, M.A., U. Hagg et al. 1996. "Dental development, dental age and tooth counts. A prospective longitudinal study of Pakistani children." *Swed Dent J* 20(1–2): 61–67.

Samaras, T.T., H. Elrick et al. 2003. "Is height related to longevity?" *Life Sci* 72(16): 1781–1802.

Saunders, E. 1837. *Teeth a Test of Age*. London: Renshaw.

Savino, F., M.F. Fissore et al. 2005. "Ghrelin, leptin and IGF-I levels in breast-fed and formula-fed infants in the first years of life." *Acta Paediatr* 94(5): 531–537.

Sawaya, A.L., P.A. Martins et al. 2004. "Long-term effects of early malnutrition on body weight regulation." *Nutr Rev* 62(7 Pt 2): S127–133.

Schlessinger, A.R., W.M. Cowan, and D.I. Gottlieb. 1975. "An autoradiographic study of the time of origin and the pattern of granule cell migration in the dentate gyrus of the rat." *J Comp Neurol* 159(2): 149–175.

Schmidt-Nielsen, K. 1975. "Scaling in biology: The consequences of size." *J Exp Zool* 194(1): 287–307.

Schneider, J.E. 2004. "Energy balance and reproduction." *Physiol Behav* 81(2): 289–317.

Schoenfeld, T., and E. Gould. 2011. "Stress, stress hormones, and adult neurogenesis." *Exp Neurol*, in press. doi: 10.1016/j.expneurol.2011.01.008.

Schubert, D., S. Humphreys, C. Baroni, and M. Cohn. 1969. "*In vitro* differentiation of a mouse neuroblastoma." *Proc Natl Acad Sci USA* 64(1): 316–323.

Schubert, D., and F. Jacob. 1970. "5-bromodeoxyuridine-induced differentiation of a neuroblastoma." *Proc Natl Acad Sci USA* 67(1): 247–254.

Schuster, D.P., K. Osei et al. 1996. "Characterization of alterations in glucose and insulin metabolism in Prader-Willi subjects." *Metabolism* 45(12): 1514–1520.

Sear, R., R. Mace, and I.A. McGregor. 2001. "Determinants of fertility in rural Gambia: An evolutionary ecological approach." Paper presented at the General Population Conference, International Union for the Scientific Study of Population, Salvador de Bahia, Brazil, August 18–24, 2001.

Sear, R., and F.W. Marlowe. 2009. "How universal are human mate choices? Size does not matter when Hadza foragers are choosing a mate." *Biol Lett* 5(5): 606–609.

Sear, R., F. Steele et al. 2002. "The effects of kin on child mortality in rural Gambia." *Demography* 39(1): 43–63.

Seiffge-Krenke, I., and T. Gelhaar. 2008. "Does successful attainment of developmental tasks lead to happiness and success in later developmental tasks? A test of Havighurst's (1948) theses." *J Adolesc* 31(1): 33–52.

Sellen, D.W. 2001. "Comparison of infant feeding patterns reported for nonindustrial populations with current recommendations." *J Nutr* 131(10): 2707–2715.

Shackleton, M., E. Quintana, E.R. Fearon, and S.J. Morrison. 2009. "Heterogeneity in cancer: Cancer stem cells versus clonal evolution." *Cell* 138(5): 822–829.

Shah, N.A., H.J. Antoine et al. 2008. "Association of androgen receptor CAG repeat polymorphism and polycystic ovary syndrome." *J Clin Endocrinol Metab* 93(5): 1939–1945.

Shanks, A.L., and G.C. Roegner. 2007. "Recruitment limitation in Dungeness crab populations is driven by variation in atmospheric forcing." *Ecology* 88(7): 1726–1737.

Shen, J.C., W.M. Rideout, and P.A. Jones. 1992. "High frequency mutagenesis by a DNA methyltransferase." *Cell* 71(7): 1073–1080.

Shors, T.J., G. Miesegaes, A. Beylin et al. 2001. "Neurogenesis in the adult is involved in the formation of trace memories." *Nature* 410(6826): 372–376.

Shors, T.J., D.A. Townsend, M. Zhao et al. 2002. "Neurogenesis may relate to some but not all types of hippocampal-dependent learning." *Hippocampus* 12(5): 578–584.

Shrimpton, R., C.G. Victora et al. 2001. "Worldwide timing of growth faltering: Implications for nutritional interventions." *Pediatrics* 107(5): E75.

Silventoinen, K., M. Bartels, D. Posthuma et al. 2007. "Genetic regulation of growth in height and weight from 3 to 12 years of age: A longitudinal study of Dutch twin children." *Twin Res Hum Genet* 10(2): 354–363.

Silventoinen, K., B. Rokholm, J. Kaprio, and T.I. Sorensen. 2010. "The genetic and environmental influences on childhood obesity: A systematic review of twin and adoption studies." *Int J Obes (Lond)* 34(1): 29–40.

Simmons, R.A. 2007. "Developmental origins of beta-cell failure in type 2 diabetes: The role of epigenetic mechanisms." *Pediatric Research* 61: 64R–67R.

Simon, P., C. Decoster et al. 1986. "Absence of human chorionic somatomammotropin during pregnancy associated with two types of gene deletion." *Hum Genet* 74(3): 235–238.

Simondon, K.B., and F. Simondon. 1998. "Mothers prolong breastfeeding of undernourished children in rural Senegal." *Int J Epidemiol* 27(3): 490–494.

Simonson, T.S., Y. Yang, C.D. Huff et al. 2010. "Genetic evidence for high-altitude adaptation in Tibet." *Science* 329: 72–75.

Sing, C.F., E. Boerwinkle, P.P. Moll, and J. Davignon. 1985. "Apolipoproteins and cardiovascular risk: Genetics and epidemiology." *Ann Biol Clin* 43: 407–417.

Sing, C.F., J.H. Stengard, and S.L.R. Kardia. 2003. "Genes, environment, and cardiovascular disease." *Arteriosclerosis Thrombosis and Vascular Biology* 23: 1190–1196.

Singer, D., and T.A. Revenson. 1998. *A Piaget Primer: How a Child Thinks*. New York: International Universities Press.

Sippell, W.G., C.J. Partsch et al. 1989. "Growth, bone maturation and pubertal development in children with the EMG-syndrome." *Clin Genet* 35(1): 20–28.

Sisk, C.L., and D.L. Foster 2004. "The neural basis of puberty and adolescence." *Nat Neurosci* 7(10): 1040–1047.

Sizonenko, P.C. 1978. "Endocrinology in preadolescents and adolescents. I. Hormonal changes during normal puberty." *Am J Dis Child* 132(7): 704–712.

Skuse, D.H. 1985. "Non-organic failure to thrive: A reappraisal." *Arch Dis Child* 60(2): 173–178.

Sladek, R., G. Rocheleau, J. Rung et al. 2007. "A genome-wide association study identifies novel risk loci for type 2 diabetes." *Nature* 445: 881–885.

Smith, B. 1992. "Life history and the evolution of human maturation." *Evolution Anthropology* 1: 134–142.

Smith, B.H. 1994. "Patterns of dental development in *Homo*, *Australopithecus*, *Pan*, and *Gorilla*." *Am J Phys Anthropol* 94(3): 307–325.

Smith, B.H., and R.L. Tompkins. 1995. "Toward a life history of the hominidae." *Annual Review of Anthropology* 24: 257–279.

Smith, J. 1964. "Group selection and kin selection." *Nature* 201(4924): 1145–1147.

Smith, P.K., B. Bogin et al. 2003. "Economic and anthropological assessments of the health of children in Maya immigrant families in the US." *Econ Hum Biol* 1(2): 145–160.

Smith, T.M., P. Tafforeau et al. 2007. "Earliest evidence of modern human life history in North African early *Homo sapiens*." *Proc Natl Acad Sci USA* 104(15): 6128–6133.

Snyder, J.S., N.S. Hong, R.J. McDonald, and J.M. Wojtowicz. 2005. "A role for adult neurogenesis in spatial long-term memory." *Neuroscience* 130(4): 843–852.

Sollars, V., X. Lu et al. 2003. "Evidence for an epigenetic mechanism by which Hsp90 acts as a capacitor for morphological evolution." *Nat Genet* 33(1): 70–74.

Soma, K.K., K. Sullivan et al. 1999. "Combined aromatase inhibitor and antiandrogen treatment decreases territorial aggression in a wild songbird during the nonbreeding season." *Gen Comp Endocrinol* 115(3): 442–453.

Sorensen, T.I., C. Holst, and A.J. Stunkard. 1992a. "Childhood body mass index—Genetic and familial environmental influences assessed in a longitudinal adoption study." *Int J Obes Relat Metab Disord* 16(9): 705–714.

Sorensen, T.I., C. Holst, A.J. Stunkard, and L.T. Skovgaard. 1992b. "Correlations of body mass index of adult adoptees and their biological and adoptive relatives." *Int J Obes Relat Metab Disord* 16(3): 227–236.

Speakman, J.R. 2008. "Thrifty genes for obesity, an attractive but flawed idea, and an alternative perspective: The 'drifty gene' hypothesis." *Int J Obes (Lond)* 32(11): 1611–1617.

Stead, J.D.H., M.E. Hurles, and A.J. Jeffreys. 2003. "Global haplotype diversity in the human insulin gene region." *Genome Research* 13: 2101–2111.

Stearns, S.C. 2000. "Life history evolution: Successes, limitations, and prospects." *Naturwissenschaften* 87(11): 476–486.

Stearns, S.C., and J.C. Koella. 2008. *Evolution in Health and Disease*, 2nd edition, Oxford University Press.

Stearns, S.C., R.M. Nesse et al. 2010. "Evolution in health and medicine Sackler colloquium: Evolutionary perspectives on health and medicine." *Proc Natl Acad Sci USA* 107(Suppl 1): 1691–1695.

Stein, A.D., H.X. Barnhart et al. 2004. "Comparison of linear growth patterns in the first three years of life across two generations in Guatemala." *Pediatrics* 113(3 Pt 1): e270–275.

Stein, A.D., F.H. Pierik et al. 2009. "Maternal exposure to the Dutch famine before conception and during pregnancy: Quality of life and depressive symptoms in adult offspring." *Epidemiology* 20(6): 909–915.

Stettler, N. 2007. "Commentary: Growing up optimally in societies undergoing the nutritional transition, public health and research challenges." *Int J Epidemiol* 36(3): 558–559.

Steudel-Numbers, K.L., T.D. Weaver, and C.M. Wall-Scheffle. 2007. "The evolution of human running: Effects of changes in lower-limb length on locomotor economy." *J Hum Evol* 53(2): 191–196.

Stöger, R. 2008. "The thrifty epigenotype: An acquired and heritable predisposition for obesity and diabetes?" *BioEssays* 30: 156–166.

Stratakis, C.A. 2003. "Genetics of adrenocortical tumors: Gatekeepers, landscapers and conductors in symphony." *Trends Endocrinol Metab* 14(9): 404–410.

Stunkard, A.J., T.I. Sorensen, C.L. Hanis et al. 1986. "An adoption study of human obesity." *New England Journal of Medicine* 314: 193–196.

Sultan, S., and S. Hasnain. 2003. "Pseudomonad strains exhibiting high level Cr(VI) resistance and Cr(VI) detoxification potential." *Bull Environ Contam Toxicol* 71(3): 473–480.

Sun, H., W. Zang, B. Zhou, L. Xu, and S. Wu. 2011. "DHEA suppresses longitudinal bone growth by acting directly at growth plate through estrogen receptors." *Endocrinology* 152: 1423–1433.

Suomi, S.J., M.L. Collins, H.F. Harlow, and G.C. Ruppenthal. 1976. "Effects of maternal and peer separations on young monkeys." *J Child Psychol Psychiatry* 17(2): 101–112.

Suzuki, M., L.S. Wright et al. 2004. "Mitotic and neurogenic effects of dehydroepiandrosterone (DHEA) on human neural stem cell cultures derived from the fetal cortex." *Proc Natl Acad Sci USA* 101(9): 3202–3207.

Suzuki, T., H. Sasano et al. 2000. "Developmental changes in steroidogenic enzymes in human postnatal adrenal cortex: Immunohistochemical studies." *Clin Endocrinol (Oxf)* 53(6): 739–747.

Szyf, M. 2009. "The early life environment and the epigenome." *Biochim Biophys Acta* 1790(9): 878–885.

Szyf, M., P. McGowan et al. 2008. "The social environment and the epigenome." *Environ Mol Mutagen* 49(1): 46–60.

Szyf, M., J. Theberge, and V. Bozovic. 1995. "Ras induces a general DNA demethylation activity in mouse embryonal P19 cells." *J Biol Chem* 270(21): 12690–12696.

Tanner, J.M. 1985. "Growth regulation and the genetics of growth." *Prog Clin Biol Res* 200: 19–32.

Tardieu, C. 1998. "Short adolescence in early hominids: Infantile and adolescent growth of the human femur." *Am J Phys Anthropol* 107(2): 163–178.

Taylor, E.N., and D.F. Denardo. 2005. "Sexual size dimorphism and growth plasticity in snakes: An experiment on the Western Diamond-backed rattlesnake (*Crotalus atrox*)." *J Exp Zoolog A Comp Exp Biol* 303(7): 598–607.

Teegarden, D., W.R. Proulx et al. 1995. "Peak bone mass in young women." *J Bone Miner Res* 10(5): 711–715.

Templeton, A.R. 2006. *Population Genetics and Microevolutionary Theory*. Hoboken, NJ: John Wiley & Sons.

Teramoto, K., K. Otoki et al. 1999. "Age-related changes in body composition of 3- to 6-year-old Japanese children." *Appl Human Sci* 18(5): 153–160.

Towne, B., S.A. Czerwinski, E.W. Demerath et al. 2005. "Heritability of age at menarche in girls from the Fels Longitudinal Study." *Am J Phys Anthropol* 128(1): 210–219.

Towne, B., S. Guo, A.F. Roche, and R.M. Siervogel. 1993. "Genetic analysis of patterns of growth in infant recumbent length." *Hum Biol* 65(6): 977–989.

Trevathan, W.R. 1996. "The evolution of bipedalism and assisted birth." *Med Anthropol Q* 10(2): 287–290.

Turnbull, J.F., and R.L. Adams. 1976. "DNA methylase: Purification from ascites cells and the effect of various DNA substrates on its activity." *Nucleic Acids Res* 3(3): 677–695.

Tutin, C. 1979. "Mating patterns and reproductive strategies in a community of wild chimpanzees (*Pan troglodytes schweinfurthii*)." *Behavioral Ecology and Sociobiology* 6: 29–38.

Uchida, A., R.G. Bribiescas et al. 2006. "Age related variation of salivary testosterone values in healthy Japanese males." *Aging Male* 9(4): 207–213.

Ulijaszek, S.J. 2001. "Body size and physical activity levels of adults on Rarotonga, the Cook Islands." *Int J Food Sci Nutr* 52(5): 453–461.

Unanue, N., R. Bazaes et al. 2007. "Adrenarche in Prader-Willi syndrome appears not related to insulin sensitivity and serum adiponectin." *Horm Res* 67(3): 152–158.

Van den Berg, S.M., and D.I. Boomsma. 2007. "The familial clustering of age at menarche in extended twin families." *Behav Genet* 37(5): 661–667.

Van der Sluis, I. 2002. *Children's Bone Health*. Rotterdam: Erasmus University.

Van Dop, C., S. Burstein et al. 1987. "Isolated gonadotropin deficiency in boys: Clinical characteristics and growth." *J Pediatr* 111(5): 684–692.

Van Goozen, S.H., W. Matthys et al. 1998. "Adrenal androgens and aggression in conduct disorder prepubertal boys and normal controls." *Biol Psychiatry* 43(2): 156–158.

Van Hoek, M., J.G. Langendonk, S.R. de Rooij et al. 2009. "Genetic variant in the IGF2BP2 gene may interact with fetal malnutrition to affect glucose metabolism." *Diabetes* 58: 1440–1444.

Van Valkenburgh, B., X. Wang et al. 2004. "Cope's rule, hypercarnivory, and extinction in North American canids." *Science* 306(5693): 101–104.

Van Weissenbruch, M.M. 2007. "Premature adrenarche, polycystic ovary syndrome and intrauterine growth retardation: Does a relationship exist?" *Curr Opin Endocrinol Diabetes Obes* 14(1): 35–40.

Vander Molen, J., L.M. Frisse, S.M. Fullerton et al. 2005. "Population genetics of CAPN10 and GPR35: Implications for the evolution of type 2 diabetes variants." *American Journal of Human Genetics* 76: 548–560.

Varela-Silva, M.I., A.R. Frisancho et al. 2007. "Behavioral, environmental, metabolic and intergenerational components of early life undernutrition leading to later obesity in developing nations and in minority groups in the U.S.A." *Coll Antropol* 31(1): 39–46.

Vaysse, P.J., and J.E. Goldman. 1990. "A clonal analysis of glial lineages in neonatal forebrain development *in vitro*." *Neuron* 5(3): 227–235.

Veldhuis, J.D., J.N. Roemmich et al. 2005. "Endocrine control of body composition in infancy, childhood, and puberty." *Endocr Rev* 26(1): 114–146.

Verhofstad, A.A. 1993. "Kinetics of adrenal medullary cells." *J Anat* 183(Pt 2): 315–326.

Vickers, M.H., P.D. Gluckman et al. 2005. "Neonatal leptin treatment reverses developmental programming." *Endocrinology* 146(10): 4211–4216.

Vilain, A., F. Apiou, B. Putrillaux, and B. Malfoy. 1998. "Assignment of candidate DNA methyltransferase gene (DNMT2) to human chromosome band 10p15.1 by in situ hybridization." *Cytogenet Cell Genet* 82(1–2): 120.

Viswanathan, S.R., G.Q. Daley, and R.I. Gregory. 2008. "Selective blockade of microRNA processing by Lin28." *Science* 320(5872): 97–100.

Wade, P.A., A. Gegonne, P.L. Jones et al. 1999. "Mi-2 complex couples DNA methylation to chromatin remodelling and histone deacetylation" [see comments]. *Nat Genet* 23(1): 62–66.

Walker, R., M. Gurven et al. 2006. "Growth rates and life histories in twenty-two small-scale societies." *Am J Hum Biol* 18(3): 295–311.

Walker, R., and M.J. Hamilton. 2008. "Life-history consequences of density dependence and the evolution of human body size." *Curr Anthropol* 49: 115–122.

Walker, R.S., M. Gurven et al. 2008. "The trade-off between number and size of offspring in humans and other primates." *Proc Biol Sci* 275(1636): 827–833.

Wang, H.S., and T. Chard. 1992. "The role of insulin-like growth factor-I and insulin-like growth factor-binding protein-1 in the control of human fetal growth." *J Endocrinol* 132(1): 11–19.

Wang, W.J., and R.H. Crompton. 2003. "Size and power required for motion with implication for the evolution of early hominids." *J Biomech* 36(9): 1237–1246.

Warren, M.P. 1983. "Effects of undernutrition on reproductive function in the human." *Endocr Rev* 4(4): 363–377.

Washburn, S.L. 1960. "Tools and human evolution." *Sci Am* 203: 63–75.

Waterland, R.A., and C. Garza. 1999. "Potential mechanisms of metabolic imprinting that lead to chronic disease." *Am J Clin Nutr* 69(2): 179–197.

Waterland, R.A., and R.L. Jirtle. 2003. "Transposable elements: Targets for early nutritional effects on epigenetic gene regulation." *Mol Cell Biol* 23(15): 5293–5300.

Waterland, R.A., and R.L. Jirtle. 2004. "Early nutrition, epigenetic changes at transposons and imprinted genes, and enhanced susceptibility to adult chronic diseases." *Nutrition* 20(1): 63–68.

Weaver, I.C., N. Cervoni, F.A. Champagne et al. 2004. "Epigenetic programming by maternal behavior." *Nat Neurosci* 7(8): 847–854.

Weaver, I.C., A.C. D'Alessio, S.E. Brown et al. 2007. "The transcription factor nerve growth factor-inducible protein a mediates epigenetic programming: Altering epigenetic marks by immediate-early genes." *J Neurosci* 27(7): 1756–1768.

Weisfeld, G. 1979. "An ethological view of human adolescence." *J Nerve Ment Dis* 167(1): 38–55.

Weisfeld, G.E., T. Czilli et al. 2003. "Possible olfaction-based mechanisms in human kin recognition and inbreeding avoidance." *J Exp Child Psychol* 85(3): 279–295.

Wells, J.C. 2003. "The thrifty phenotype hypothesis: Thrifty offspring or thrifty mother?" *J Theor Biol* 221(1): 143–161.

Wells, J.C. 2009. "Thrift: A guide to thrifty genes, thrifty phenotypes and thrifty norms." *Int J Obes (Lond)* 33(12): 1331–1338.

West, G.B., J.H. Brown et al. 1997. "A general model for the origin of allometric scaling laws in biology." *Science* 276(5309): 122–126.

West-Eberhard, M.J. 2005. "Phenotypic accommodation: Adaptive innovation due to developmental plasticity." *J Exp Zoolog B Mol Dev Evol* 304(6): 610–618.

White, S. 1996. "The child's entry into the age of reason." In: *The Five To Seven Year Shift: The Age of Reason and Responsibility*," edited by A. Sameroff and M.M. Haith, 17–32. Chicago: University of Chicago Press.

White, T.D., B. Asfaw, Y. Beyene et al. 2009. "*Ardipithecus ramidus* and the paleobiology of early hominids." *Science* 326(5949): 75–86.

Whitelaw, N.C., and E. Whitelaw. 2006. "How lifetimes shape epigenotype within and across generations." *Hum Mol Genet* 15(Suppl 2): R131–137.

Whiten, A., and A. Mesoudi. 2008. "Establishing an experimental science of culture: Animal social diffusion experiments." *Philos Trans R Soc Lond B Biol Sci* 363(1509): 3477–3488.

WHO Working Group on Infant Growth. 1995. "An evaluation of infant growth: The use and interpretation of anthropometry in infants." *Bull World Health Organ* 73(2): 165–174.

Widdowson, E.M., and R.A. McCance. 1975. "A review: New thoughts on growth." *Pediatr Res* 9(3): 154–156.

Wilks, A., M. Seldran, and J.P. Jost. 1984. "An estrogen-dependent demethylation at the 5' end of the chicken vitellogenin gene is independent of DNA synthesis." *Nucleic Acids Res* 12(2): 1163–1177.

Williams, G. 1957. "Pleiotropy, natural selection, and the evolution of senescence." *Evolution* 11(4): 398–411.

Wilson, R.S. 1979. "Twin growth: Initial deficit, recovery, and trends in concordance from birth to nine years." *Ann Hum Biol* 6(3): 205–220.

Wingfield, J.C., J. Jacobs et al. 1997. "Ecological constraints and the evolution of hormone-behavior interrelationships." *Ann NY Acad Sci* 807: 22–41.

Winocur, G., J.M. Wojtowicz, M. Sekeres et al. 2006. "Inhibition of neurogenesis interferes with hippocampus-dependent memory function." *Hippocampus* 16(3): 296–304.

Winter, J.S., and C. Faiman 1973. "Pituitary-gonadal relations in female children and adolescents." *Pediatr Res* 7(12): 948–953.

Wit, J.M., and H. van Unen. 1992. "Growth of infants with neonatal growth hormone deficiency." *Arch Dis Child* 67(7): 920–924.

Witchel, S.F., R. Smith et al. 2001. "Candidate gene analysis in premature pubarche and adolescent hyperandrogenism." *Fertil Steril* 75(4): 724–730.

Wolff, G.L., R.L. Kodell et al. 1998. "Maternal epigenetics and methyl supplements affect agouti gene expression in Avy/a mice." *FASEB J* 12(11): 949–957.

Wollmann, H.A., T. Kirchner et al. 1995. "Growth and symptoms in Silver-Russell syndrome: Review on the basis of 386 patients." *Eur J Pediatr* 154(12): 958–968.

Woo, J.G., M.L. Guerrero et al. 2009. "Human milk adiponectin is associated with infant growth in two independent cohorts." *Breastfeed Med* 4(2): 101–109.

Xu, X., W. Wang et al. 2002. "Longitudinal growth during infancy and childhood in children from Shanghai: Predictors and consequences of the age at onset of the childhood phase of growth." *Pediatr Res* 51(3): 377–385.

Yeh, P.J., and T.D. Price. 2004. "Adaptive phenotypic plasticity and the successful colonization of a novel environment." *Am Nat* 164(4): 531–542.

Yoder, J.A., C.P. Walsh, and T.H. Bestor. 1997. "Cytosine methylation and the ecology of intragenomic parasites" [see comments]. *Trends Genet* 13(8): 335–340.

Yu, J., M.A. Vodyanik, K. Smuga-Otto et al. 2007. "Induced pluripotent stem cell lines derived from human somatic cells." *Science* 318(5858): 1917–1920.

Zadik, Z., S.A. Chalew et al. 1985. "The influence of age on the 24-hour integrated concentration of growth hormone in normal individuals." *J Clin Endocrinol Metab* 60(3): 513–516.

Zahavi, A. 1977. "The cost of honesty (further remarks on the handicap principle)." *J Theor Biol* 67(3): 603–605.

Zareparsi, S., R. Camicioli, G. Sexton et al. 2002. "Age at onset of Parkinson disease and apolipoprotein E genotypes." *American Journal of Medical Genetics* 107: 156–161.

Zeisel, S.H. 2009. "Epigenetic mechanisms for nutrition determinants of later health outcomes." *Am J Clin Nutr* 89(5): 1488S–1493S.

Zemel, B.S., and S.H. Katz. 1986. "The contribution of adrenal and gonadal androgens to the growth in height of adolescent males." *Am J Phys Anthropol* 71(4): 459–466.

Zhang, R.L., Z.G. Zhang, and M. Chopp. 2008. "Ischemic stroke and neurogenesis in the subventricular zone." *Neuropharmacology* 55(3): 345–352.

Zhu, H., S. Shah, N. Shyh-Chang et al. 2010. "Lin28a transgenic mice manifest size and puberty phenotypes identified in human genetic association studies." *Nat Genet* 42(7): 626–630.

Zingg, J.M., J.C. Shen, and P.A. Jones. 1998. "Enzyme-mediated cytosine deamination by the bacterial methyltransferase M.MspI." *Biochem J* 332(Pt 1): 223–230.

Zuckerman-Levin, N., and Z. Hochberg. 2007. "Delayed infancy-childhood spurt (DICS) in SGA children with no catch up growth." *Horm Res* 68(Suppl 1): 167.

Zverev, Y., and J. Chisi. 2004. "Anthropometric indices in rural Malawians aged 45–75 years." *Ann Hum Biol* 31(1): 29–37.

INDEX

Evo-Devo of Child Growth: Treatise on Child Growth and Human Evolution, First Edition.
Ze'ev Hochberg.
© 2012 Wiley-Blackwell. Published 2012 by John Wiley & Sons, Inc.